English for the English

ENGLISH FOR THE ENGLISH

CAMBRIDGE UNIVERSITY PRESS
C F. CLAY, Manager
LONDON   FETTER LANE, E.C. 4

NEW YORK   THE MACMILLAN CO
BOMBAY
CALCUTTA   MACMILLAN AND CO., Ltd.
MADRAS
TORONTO   THE MACMILLAN CO OF
CANADA, Ltd.
TOKYO . MARUZEN-KABUSHIKI-KAISHA

English for the English

A CHAPTER ON NATIONAL EDUCATION

BY

GEORGE SAMPSON

Hon. M.A. Cambridge
st john's college

I love Rome, but London better, I favour Italy, but England more,
I honour the Latin, but I worship the English

RICHARD MULCASTER (1582)

To reconcile man as he is to the world as it is, to preserve and improve
all that is good, and destroy or alleviate all that is evil, in physical and
moral nature—have been the hope and aim of the greatest teachers
and ornaments of our species

THOMAS LOVE PEACOCK (1818)

CAMBRIDGE
AT THE UNIVERSITY PRESS
1922

TO

M. AND D.

First Edition 1921,
Reprinted 1921, 1922

PREFACE

THE main subject of this essay is the education of boys and girls up to the age of fourteen or fifteen. The children concerned will be found (1) in the Elementary Schools, (2) in the Preparatory Schools, (3) in the lower forms of the Secondary Schools. My experience having lain with boys in the first of these groups, I have naturally made them the theme of my discourse; but most of the matter is equally applicable to the other two groups. Indeed, I hope that teachers of all grades and subjects will find something here to interest them. What I hope for most of all is to find a few readers among the general public, all of whom, remember, as citizens, tax-payers, rate-payers or parents, are vitally concerned in education. There is a Navy League to stimulate general interest in the Navy and make the Estimates seem worth while. Will someone found an Education League?

I have wandered now and then from my specific ground of juvenile instruction into wider fields of education, but I do not apologise and I do not repent. Digression is sometimes the better part of travel.

As I am unwilling to ask for more than one revolution at a time, I have taken the present organisation (or rather no organisation) of schools as it is, and I have said nothing about such interesting topics as: (1) whether all education after the age of twelve ought not to be considered as secondary education, but not necessarily of one kind, (2) whether Central Schools, Higher Grade Schools, Higher Standard Schools, Continuation Schools and all other devices of Authority for evading the provision of a proper scheme of secondary education ought not to be abolished; (3) whether the present elaborate isolation of Secondary Schools from the main elementary high road, as shown by the difference in scale of building and equipment, in status and pay of teachers, in size of classes, and even in such trifles as length of school day, incidence of term, and duration of holiday is not a display of diseased class consciousness and, therefore, unnatural and provocative, (4) whether the official depreciation of elementary education as an inferior branch of public service implying social and intellectual inferiority in the teachers is not a real cause of the fatal shortage of teachers, especially of men teachers; (5) whether training colleges are not now ripe for extinction instead of extension, their work in the education of prospective teachers being done by the universities, to which it properly belongs, and whether the people who imagine that teachers are taught how to teach in the training colleges should not be reminded that there is only one place in which a person can learn how to teach, and that is the school—that, in fact, one learns how to teach by teaching.

All these are modest questions which I shall not discuss at the moment. The revolution I am now asking for is a bloodless revolution, involving no more than a conversion or change of heart in teachers and officials, and the destruction of most existing time-tables and syllabuses. In fact, I am asking for nothing new at all. I merely propose that certain aspects of school work, now dimly recognised as desirable, should be clearly recognised as the most important of all, and made the chief charge upon the available time and energy. As an inevitable corollary (time being as short as ever) some of the school subjects now receiving most attention will have to receive less Briefly, what I urge in these pages is that elementary education must be elementary— that it should aim at beginning something, and not at finishing everything. I hasten to assure the suspicious tax-payer who buttons up his pockets at the sound of the word education that the reform I propose will not add a farthing to national expenditure, but will give better value for money spent.

These pages resume the substance and sometimes the actual words of various articles, addresses and memoranda of mine, ranging over a period of several years. Any authority they possess must derive from the fact that they present, not the speculations of a theorist, but the convictions of a teacher who has been engaged in elementary school work for twenty-five years, and who feels more certain with every added year that the present elementary system is a failure and needs re-orientation. I believe that the great purpose of education is not to make people *know* something but to make people *be* something. I believe that purpose is not at present fulfilled by our schools.

I believe that the recommended interest of teachers in the

'science' of education—in 'psycho-analysis' (imported from Germany) in 'tests of intelligence' (imported from France) and in 'experimental psychology' (imported from America)—means excessive concern with the heads of children and no concern for their souls.

I believe that recent gospels of anarchy urging that children must never be instructed or restrained, but that they must always be allowed to do just what they like, when they like and how they like, are false to the purpose of education, which is to prepare mankind, not merely to live, but to live together in human fellowship and reasonable subordination, here and now, in the very world that is the world of all of us.

I believe that the course now followed in schools is in substance unpractical and fanciful, that it does not give the public what the public wants. I believe that this is true, not only of the elementary schools, but of all schools that deal with the education of 'juveniles.' If I may, without irreverence, parody a question sacred to a beautiful legend, I want to ask briefly and pointedly, Quo vadis, dominie? What exactly are you about? Have you any clear notion of what you are trying to do for our boys and girls? Adapting other phrases (not at all sacred, but not without beauty), I urge in this essay that all English children, whatever their schools may be called, shall have a practical education that will fit them for their station in life. What their station in life really is, is a question we have scarcely yet begun to ask ourselves The future of the nation depends upon how we answer it

G. S.

BARNES,
July 14, 1921.

CONTENTS

I

Preliminary

LORDE GOD, howe many good and clene wittes of children
be nowe a dayes perisshed by ignorant schole maisters.
SIR THOMAS ELYOT.

HALF a century ago, the State decreed that all its children
should receive some sort of instruction. For those who
could pay, there already existed many schools, public and private,
endowed by deceased benefactors or conducted for personal profit.
These ranged from Eton and Winchester to Miss Pinkerton's
Academy on the Chiswick Mall. For the poorer classes there also
existed schools, public and private, supported by benevolent funds
or conducted for personal profit. These ranged from the National
and British Schools to the classes conducted by Mr Wopsle's great-
aunt. It was the privilege of the English parent to choose whether
his children should be instructed or not—whether his son should
be as tutored as Tom Jones or as unlettered as Tony Lumpkin.
The Education Act of 1870 abolished this option. From that date
the entire youth of England had to be subjected to a recognised
form of instruction. This was called education.

The half-century since the passing of this Act has been a
wonderful period of experiment, adventure and speculation. A
comparison of the educational activities of 1871 with those of
1921 discloses an amazing growth, not merely in the quantity,
but in the spirit and the rate of development. A period of contented stagnation has been followed by a period of excitement.
Looking back we might say that the first part of those fifty years
was spent in an unconscious demonstration of what education is
not, and the second part in eager attempts to decide what education is

To this development the spirit of the age in science has contri-
buted. The seventeenth century was the century of mathematics,

the eighteenth century the century of chemistry, the nineteenth century the century of biology. The twentieth century seems likely to be the century of psychology—not the old psychology of supposition, but the new psychology of practical investigation. The schools are wonderful laboratories for experiment, and teachers have already been told by persons of official importance that they will not be respected unless they make definite contributions to educational science. The art of teaching, it would appear, is not respectable. That an art is promoted by claiming to be a science is a view entirely and delightfully official. It might be called the traditional Front Bench view of all the arts. Certainly the scientific attitude in schools is leading to a new interest in children. There are already societies for the study of infancy, youth and adolescence. The juvenile instincts, formerly assumed to be wicked, are now respected and cultivated as materials. Soon, no doubt, we shall go even further. Budding psychologists will walk the schools as medical students walk the hospitals, and children will be as useful to the educationist as dogs are to the vivisectionist. They will be inoculated with doses of knowledge and tested for reactions; their minds will be calibrated periodically with millimetric exactness, and their abilities neatly reduced to succinct tables and beautifully sinuous curves. The psychologist will rejoice, and so will the routine-loving official, whose cardiac system leaps up when he beholds statistics from on high. And thus the juvenile mind will be docketed and disposed of, and education comfortably removed from the troublesome world of feeling.

Experimental work may be very valuable, and I am not so stupid as to oppose or decry it, but it is not everybody's work, and it must not be the first charge upon the energies of teachers. I want teachers to remember that they are first of all healers and not vivisectionists. I want them to see clearly that laboratory work in school is not education, and that to test a child's mind is not to teach it. Dogs are not really improved by vivisection even if the mind of the vivisector is. Teachers may, if they will, conduct a series of very useful experiments in school, but that is not what they are there for. Teachers go to school to practice the art of teaching, not to pursue the science of education. What the teacher has to consider is not the minds he can measure but the souls he can save. Nothing is easier than to neglect children for the pursuit of neurograms. Neurograms are so much easier to manage Psychology can and should assist the teacher, but it must not obsess the teacher. Let me put it this way: If a hungry child came to you, you might find him an interesting field for a study of the phenomena of starvation. You could compile illuminating graphs of his reaction to various stimuli, and you might even work out mathematically how long it would take him to die, and check the result by experiment. You might do all these scientific things, but the obvious human thing to do would be to give him something to eat. If teachers abandon the art of teaching for the science of education, they may compile some ingenious and valuable statistics; but while the shepherds are thus dallying with the delights of mathematics, the hungry sheep look up, and are not fed.

One danger of the present passion for science has already shown its head. Psychology is becoming the hand-maid of educational reactionaries as chemistry became the hand-maid of the war-lords. Delight in measured results means a demand for results that can be measured. The tables and curves of the psychologist are taking us back to examinations and a prescribed minimum standard of attainment—that is, just to the old demand that has already ruined the past and imperilled the future of elementary education. I say 'ruined,' because, with all its good intentions and its great achievements, our system of elementary education is a failure. This is the test. An educational system that is not a failure will produce an educated population. The present system obviously does not do that; and it will continue not to do that, if the energy of teachers, having been diverted from one kind of examination, is directed towards another. Elementary education has failed because we have thought too much of the children's heads and not enough of their hearts. Hearts are still out of fashion in school. In spite of its name psychology has nothing to do with the soul.

Much, indeed, has been accomplished in the past half-century of national education; but one great need has been forgotten. We have tried to educate the children: we have scarcely even tried to educate the public. Before educational progress is possible the public must be taught the meaning of education; and to this work our official leaders and spokesmen should turn their strongest efforts. The national mind must be got to see that education is a spirit and not a substance Education is not something of which

we must acquire a certain quantity and can then be relieved for \
ever. There is no end to education. There are no 'finishing schools,' \
except in the worst sense, and most of them are that. We might
as well try to get in our 'teens the minimum of righteousness that
will admit us to heaven and consider that we are then 'finished'
with religion. Education is initiation, not apprenticeship. It has
nothing to do with trade, business or livelihood; it has no connection with rate of wages or increase of pay. Its scale is not the
material scale of the market. Education is a preparation for life,
not merely for a livelihood, for living not for a living. Its aim is
to make men and women, not 'hands.' In moments of expansive-
ness we may admit this as an idea or ideal; but we deny it in our
practice. If we are really sincere, this ideal must inspire not
merely our educational talk, but our educational deeds.

A very admirable, hard-working lady came one day to a London
elementary school on Care Committee business, and found that
the 'leavers' she wanted to interview had gone with their class
to a performance of *Twelfth Night*. "Of course," she said, quite
pleasantly, "it is very nice for the boys to go to the theatre, but
Shakespeare won't help them to earn their living." This is pro-
foundly true. Shakespeare will not help anyone to earn a living,
not even a modern actor-manager. Shakespeare is quite useless, as
useless as Beauty and Love and Joy and Laughter, all of which
many reputable persons would like to banish from the schools of
the poor. Yet it is in beauty and love and joy and laughter that we
must find the way of speaking to the soul—the soul, that does not
appear in the statistics and is therefore always left out of account.
By a dreadful inversion we conceive the proper progress of a poor
child to be from the Purgatory of school to the Hell of labour.
Happiness in the class-room is still regarded with suspicion. A
director of a large business house that had its own Day Continuation School came and watched a class of his girls reading *A
Midsummer Night's Dream* with an enthusiastic teacher. At the
end he asked, "Do they really get any good from the school or
do they just amuse themselves like this?" Teachers are no better
than the rest in their view of education Their hearts are set upon
what they call 'practical subjects,' and when they appeal to
children to continue their education in evening schools, the
strongest argument they adduce is a possible improvement of

position. And so children think they learn in order to earn, and cannot imagine any other purpose in learning.

Those who saw J. M. Barrie's *Mary Rose* will remember the astonishment with which Mary Rose and her husband Simon Blake discover that their Highland gillie is a prospective minister, studying at the University of Aberdeen, and earning from English tourists in the vacation enough to support him during term. An even greater surprise awaits them:

Mary Rose. Is your father a crofter in the village?
Cameron. He is, Ma'am, when he is not at the University of Aberdeen.
Simon. Great snakes! Does he go there too?
Cameron. He does so. We share a very small room between us.
Simon. Father and son! Is he going into the ministry too?
Cameron. Such is not his purpose. When he has taken his degree he will
    return and be a crofter again
Simon In that case I don't see what he is getting out of it.
Cameron. He is getting the grandest thing in the world out of it. He is
    getting education.
Simon. You make me feel small.

This is excellent; but it should be pointed out to Barrie that Simon Blake represents, not exactly the English view, but what a charitable Scotsman imagines the English view would be. That Simon should say "I don't see what he is getting out of it," is quite English; but that, being told the elderly student would get out of it the greatest thing in the world, education, he should reply "I feel small," is not English. It is merely what a Scotsman would expect an Englishman to say. The spectacle of disinterested education would not make an Englishman feel small; it would make him feel contemptuous or hilarious (Here, as an illustration of our national disinterestedness, the congregation will sing Hymn 365 from *Hymns Ancient and Modern*:

> *mf.* Whatever, Lord, we lend to Thee
> *cr.* Repaid a thousandfold will be,
> *f.* Then gladly will we lend to Thee,
>   Who givest all.

Observe the financial ecstasy of the musical directions. The sentiment may be Ancient; it is certainly Modern. That being the national view of religion, we must not be too hopeful about disinterested education!)

Our English view of education is that, like honesty, it is the best policy. It is an investment—a bit risky, but worth trying. You put something into education and you may get a substantial return. If you don't, education is a fraud. That is both the collective view and the individual view. No public money is so grudged as the money spent on education, and it is the expenditure that comes most closely home to the bosom of the citizen, because much of the cost is borne by the local rates. I suggest that the idealists who want to abolish war should start a world movement for putting all the armies and navies on the rates. Militarism would soon be unpopular when it clearly worked out at so much in the pound every quarter.

Let us repeat—and we shall say it again before the end of this chapter—that education is not a 'business proposition.' It is like pure religion and undefiled, without a cash value. It is not an apprenticeship to a trade, and it may have no relation to success in a chosen calling. Let me take a simple example. When I go home at the end of a day's work, I show my season ticket to a man at the door of a Tube lift. I reach the platform, and, when the train arrives, the gates are opened and closed by a second man, and the train goes off, started and stopped by a third man. At the end of the train journey I show my ticket to a fourth man, and I go into the street, get on a 'bus, where a fifth man collects my fare and gives me a ticket as the 'bus careers along, guided by a sixth man on the front seat. Probably before the end of the journey I shall have to produce my ticket for inspection by a seventh man. Consider the lift-men, gate-men, motor-men, conductors; consider the steady growth of occupation in the direction of tasks like theirs, in which there is the almost mechanical repetition of almost mechanical actions! Modern mass-production does not require educated workmen; it scarcely needs even intelligent workmen. How can it be pretended that education has any specific application to tasks in which there is no need for intelligence? The lift-man would work his switch no worse if he were quite illiterate and no better if he were a doctor of science. It is not as a lift-man that he is worth educating, but as a man. That is what the nation must be persuaded to see If the nation exclaims, "What! all this education, and he only a lift-man!" it is uttering wickedness and stupidity: if the lift-man exclaims,

"What! all this education, and me only a lift-man!" he is uttering wickedness and doubtful grammar. The lift-man must be told that if he studied in his leisure and took a science degree he would not therefore become a director. The politicians must be told that when they denounce the education of children above their station they are talking blasphemy. You cannot educate children above their station, for you are educating men and women; and in this world there is no higher station.

The relation of technical work to education is an interesting subject. The position is all the clearer now that a belief in formal training, or general 'mental gymnastic,' has been abandoned. Stated roughly, the old idea was that almost any subject would do to train the mind upon, provided that it was difficult enough— and some martinets would add, disagreeable enough. It was believed that the traditional classical curriculum gave a general mental alertness and strength that was valid for all purposes of life; skill in mathematics produced a special intellectual acuteness applicable to most circumstances; Euclid cultivated the general reasoning power; and so forth. Investigation has disproved these claims—and, indeed, for the plain man they hardly needed dis- proving. It is quite obvious that the great majority of classically educated persons do not possess any remarkable alertness or strength or breadth of mind; that persons addicted to the solution of arithmetical puzzles are usually intolerable dullards; that champion chess players, so far from being remarkable for a high degree of foresight, and what may be called the practical strategy and tactics of life, are among the stupidest of mankind. If you labour to become specially good at Algebra (for instance) you do not get a general mental training that will make you good at Botany. If you want to be proficient in Botany, you must study Botany and not Algebra. Our railways are built by engineers and surveyors, our bodies are tended by physicians, surgeons and den- tists, our music is provided by composers and executants who are immensely skilled in their own calling, but who may be, and often are, persons who cannot in the broad sense, be called educated. Their elaborate training has carried over nothing at all to their general education What have they in common? Remove their 'shop' (and perhaps their golf) and they are ignorant and in- articulate creatures, uninteresting and uninterested. Vocational studies may contribute very largely to education, but intensive work at any professional or vocational subject does not in itself make an educated person. However, we need not here discuss vocational studies, for it ought to be plain to everybody that vocational studies and professional training should have no place in the education of children at the elementary school age. In the preface to *Unto This Last*, Ruskin, writing eight years before the passing of the Elementary Education Act of 1870, demanded public schools in which a child should "imperatively be taught, with the best skill of teaching that the country could produce, the following three things:

(a) the laws of health and the exercises enjoined by them;
(b) habits of gentleness and justice;
(c) and the calling by which he is to live "

It will be seen in the sequel how far there is still need for the first two prescriptions There is none for the third. The only 'calling' that should be taught in school is the state of manhood, to which we are all called. Every occupation, every variety of occupation, has its own peculiarities, which must be acquired in the appropriate place, and cannot be acquired at school. Ruskin, like most other critics of education, would have been all the better for a few years' work as a teacher in a large elementary school. Here, for instance, is a school of four hundred and fifty boys. Ask them what calling they intend to pursue, and they will unanimously reply 'Engineering.' Shall we, therefore, resort to a curriculum of universal engineering? We had better not, because very few, perhaps not one, of all the engineering aspirants will ever see the inside of an engineering shop. The fallacy of the Ruskinian prescription lies in the supposition that a boy chooses a calling. What usually happens is that a calling chooses a boy. Here are nineteen boys leaving school at the end of a term. Four of them are going to secondary schools. Three of those who wanted to be engineers are going straight into an office, because a certain well-known firm has written to ask for three boys Had an engineering firm written for three boys they would have gone into engineering They are three good boys, and they will be just as good clerks as they would have been engineers. Two of the fifteen are going into the Post-office as messengers. One is going into a hair-dresser's shop—a useful, and, in fact, an indispensable institution; but we can hardly teach the calling of Figaro in school, nor indeed, can I, at the moment, discover any school subject that has vocational bearings in that direction. Another of the fifteen is taken on by a picture-frame maker, and doubtless his manual training lessons will be of use to him. Another is going to work with a sign-writer, another is going into a house-decorator's shop; several cannot find jobs at all, and some of them will inevitably find their way on to a tradesman's tricycle. All this variety is repeated with the leaving of every batch. Clearly, we should have an exciting time in town schools if we ventured far in the direction of the calling by which the boys will live! Nor is the problem really different in the districts of one occupation. Think of the schools in our mining villages. The boys are the sons of miners and the brothers of miners. They are born into the atmosphere of the mine, they grow up in the shadow of the mine, and in due course the mine swallows them up. No sane person can really believe that it is the duty of the schools to underline the obvious, and tune the minds of these children for ever to all the grim and dusky circumstances that encompass them. To put the matter quite plainly, I deny utterly that it is the business of the elementary school in a mining village to teach the boys to be miners. I assert, on the contrary, that the elementary school in a mining village must be conducted without any reference whatever to coals or mines or managers or directors or shareholders.

And let us look at another picture. Here is a model factory employing girls taken straight from the elementary school. One girl spends the whole of her day in putting sheets of tin plate into the slot of a printing machine; another spends the whole of her day in wiping the printed tin sheets with an oily rag; another spends her day in putting the printed sheets into a stamping machine; another spends her day in collecting the scrap tin that comes from the stamping machine; another spends her day in collecting the tin boxes stamped out by the machine; another spends her day in placing the tins in position under an automatic filler, another spends her day in putting lids on the filled tins, another spends her day in packing the filled tins in cartons; and so on, and as a rule not one of these girls ever does anything else in all her factory life. Consider! Here are girls who daily for five or ten or a dozen years do nothing else but put tin lids on boxes; and all over the country there are hundreds of thousands—adolescents and adults of both sexes—whose wage-earning life is spent in tasks just as brainless, just as maddeningly mechanical. To speak in abstractions is always impressive and always easy. It sounds almost morally convincing to say that "education must prepare a child for the calling by which he is to live." It sounds almost scientifically convincing to say that "education should be given a vocational bias with reference to the specific character of the local industries." But let us leave abstractions, moral and scientific, and speak in facts. Is anyone prepared to maintain that the purpose of elementary education is to teach boys to hew out coal and girls to put lids on boxes? I am prepared to maintain, and, indeed, do maintain, without any reservations or perhapses, that it is the purpose of education, not to prepare children *for* their occupations, but to prepare children *against* their occupations. I asked if anyone thought the purpose of education was to teach girls how to put lids on boxes. The Ruskinian idealist would, of course, reply that the spectacle of girls putting lids on boxes is abhorrent to God and a blot on the face of Nature, and, therefore, something not to be prepared for, but to be utterly wiped out. All of which may be true; but, however sincerely we desire to reform the world, we must, for immediate purposes, take the world as it is; and this happens to be, just now, a world in which girls are required to put lids on boxes. It is, fortunately, also a world in which those girls are compelled to go to school for several years. The practical view will therefore embrace both these facts, and regard this as a world in which girls have to be prepared, not, indeed, for putting lids on boxes, but for a life of which putting lids on boxes will be a large part. We must really get out of the habit of talking as if education were the preparation of children merely for that part of their life which does not belong to them, as if they, as reasonable, living beings, had no existence at all. The attempt to relate elementary education with wage-earning seems to me not merely impracticable, but wrong in principle. There are now schools with a 'vocational bias' into which children of eleven are drafted. All we need say is that any serious attempt to give children of eleven a vocational education is a crime, and any pretended attempt a sham. Till the age of at least fourteen it is our duty to educate a child, not train a 'hand.' The child will become a hand quite soon enough: the schools need not hasten the process

What is vitally wrong in the curriculum of such schools is not merely the vocational inclusions, but the vocational exclusions. It is bad enough that children of this age should actually receive regular protracted instruction in shorthand and type-writing, both of which are purely mechanical activities, best acquired in a short period of intensive practice; but worse than such inclusions, worse even than the bias of mere utility given to a subject of delight like drawing, are the deprivations, the exclusion of everything that does not pertain to vocation. Thus, few school activities are more humanising than music. Music on its practical and receptive side —the singing of songs and the listening to pieces played—is of high importance in education. I have known classes of wild and boisterous factory girls in a continuation school soften into docility and disciplined eagerness as a result of their music, and there was no lesson they asked for so eagerly or missed so regretfully. Now it would be interesting to know in how many vocational or junior technical schools music is taken seriously as a necessary part of school life. The mere appearance of the name on a time-table, of course, means little. The important question is whether music is actually treated by all in the school as an essential element in the education of children, or whether it is not usually left to cling to the skirts of unhappy chance, as something unworthy of effort compared with shorthand, type-writing and the manipulation of the slide rule. A school without music should be impossible.

It is, I think, indisputable that most people think of an elementary school as something quite different in purpose from a public school. The difference may be put thus: Harrow is allowed to make men: Hoxton has to make hands. From the elementary schools employers expect to receive a steady supply of acquiescent and well-equipped employees. If the supply does not reach expectation, the elementary schools are denounced. Remember that boys leave the elementary school to go to work at just the age when more fortunate boys leave the preparatory schools to go to public schools. That is to say, the elementary schoolboy has done with school when the public schoolboy is just beginning. Many of them are physically mere children at fourteen, and, pray, what is to be expected from the mind at that age? Yet I have regularly received applications from employers for boys *qualified in short-hand and typewriting.* That is, employers want the elementary schools to give them clerks able to do an adult's work at a juvenile's wage, and they denounce the schools because their amiable desires are not satisfied.

Unfortunately, many persons in authority are inclined to take the alleged 'practical' point of view, and to test the efficiency of a school by suddenly asking the children to multiply £24. 3s. 7½d. by 29. Authority cannot measure the amount of personal refinement or general development Tom Brown has acquired during his six or seven years at school, so Authority takes no notice of that; but Authority can easily measure Tom Brown's failure to multiply £24. 3s. 7½d. by 29, and can even publish the statistics. Tom Brown may have been changed (I speak of an actual instance) from a dirty little street boy, truanting and cadging, to the steady lad who now proudly officiates as prefect; but Tom Brown's ability to multiply £24. 3s. 7½d. by 29 is very precarious, and so Tom Brown is denounced as one of the 'failures' of our costly elementary schools, when he is really one of the successes. His teacher is suspected as a slacker, and his teacher, therefore, grows to think it wiser to conform to the world. After all, it is much easier to make Tom Brown multiply £24. 3s. 7½d. by 29 than to make him a decent fellow. The elementary school-master concentrates on making his boys pass the scrutiny of Authority and the Practical Business Man, just as the preparatory school-master concentrates on making his boys pass the scholarship and entrance examinations of the public schools The total result is that a few (who would prosper in any circumstances) manage to do well; but the many, the average boys, remain half crammed and almost entirely uneducated.

What is the purpose of education, to make Tom Brown increase in grace or multiply in arithmetic? The schools have hesitated, have said one thing and done another. The old elementary school had no choice. It was told to make Tom Brown multiply, and it made him multiply with extraordinary success; but it did not make him grow in grace, and did not pretend to. Now that fuller liberty has come, the elementary school is not quite sure which it ought to do, and so it does neither quite well.

Our schools for the young have failed because they have tried for obvious and immediate results instead of trying to lay the foundations of a humane education; and so the melancholy fact remains that, after fifty years of compulsory schooling, the Englishman is still uneducated. That, at least, is a 'result' that cannot be challenged! Let the schools, therefore, and especially the elementary schools, abandon their so-called 'practical' aims and their concern with some vague and disagreeable 'livelihood' awaiting the pupils, and, leaving results to time and natural development, turn their chief attention to the ideal embodied in that which, because its aim is to make men and women and not machines, we call the Humanities By a strangely benevolent coincidence, the elementary schools, in which the need for a purely personal and humane form of education is most pressing, have by far the greatest freedom in shaping instruction to the wants of the pupils, as they alone, of all educational institutions, are free from the binding tyranny of external examinations. What provides a liberal education may be matter of controversy between the Franciscans of classics and the Dominicans of science; but the truth is that it is a matter of attitude rather than of subject—that a man may be liberally educated in science and illiberally learned in classics. The commercial attitude, the attempt to touch an early and assured dividend, is fatal to education by whatever means pursued. In education there are no debenture subjects, there are no short cuts to culture. Fortunately, however, the controversies of extremists do not, or should not, affect the early stages of education; —I say 'should not,' because even here it is difficult to escape the classical extremist who insists that classical scholarship will perish from this land unless children imbibe the ancient tongues with their mothers' milk. The infliction of Latin grammar upon infants of seven seems to me not so much a subject for educational discussion as a case for the N.S.P.C C Whatever form the later education of boys and girls may take, whatever the special subject or aspect of study they elect to pursue, whatever the nature of the livelihood that actually awaits them at fourteen or sixteen or eighteen or twenty-one, they must all be able to speak, to read, to write, because speaking, reading, and writing are the means of human intercourse, of communion between man and his fellows. The inarticulate person is cut off from his kind or fatally limited to a communion of sullen contact with the equally inarticulate. Before the English child can awaken to any creative fullness of life he must become proficient in the use of his native tongue, the universal tool of all callings and of all conditions. To assert the importance of English for the English, to offer both a plea and a programme for a really practical education is the purpose of the following pages.

II

A Plea

IT is only by amusing oneself that one can learn....The whole art of teaching is only the art of awakening the natural curiosity of young minds for the purpose of satisfying it afterwards; and curiosity itself can be vivid and wholesome only in proportion as the mind is contented and happy. Those acquirements crammed by force into the minds of children simply clog and stifle intelligence. In order that knowledge be properly digested, it must have been swallowed with a good appetite. If that child were intrusted to my care, I should make her—not a learned woman, for I would look to her future happiness only—but a child full of bright intelligence and full of life, in whom everything beautiful in art or nature would awaken some gentle responsive thrill I would teach her to live in sympathy with all that is beautiful— comely landscapes, the ideal scenes of poetry and history, the emotional charm of noble music. I would make lovable to her everything I would wish her to love.

ANATOLE FRANCE.

THERE is a science of education, but education is not a science. No one is entitled to lay down principles of teaching as if they were laws of mechanics. The utmost a teacher can do is to examine his beliefs and his experience (including, of course, his observation of the pupils) and see what guidance they afford him. I propose to discuss here the place of our language and literature in the early stages of education. I shall not discuss methods of teaching, and may scarcely even refer to them. It should be said at once that the methods of teaching English have very perceptibly changed for the better; but it cannot be said that the place of English in education is clearly seen by many who teach it well. At this stage of our national education what matters is the faith, not the works. We have (so to speak) to undergo conversion, not to practise new austerities. It is more important to go to Damascus than to go to Canossa.

In considering the place of English in our scheme of national education one thinks chiefly and immediately of the elementary schools. Indeed, they should be thought of first and nearly all the time, for they are not only important in themselves, but what is true of them will be largely true of all other schools containing children of the elementary age. I will begin, therefore, by stating two propositions, the first of which needs no discussion·

(1) That the elementary schools are by far the most important schools in the country; and (2) that English is by far the most important subject in the elementary schools. These two propositions depend upon an assumption that the purpose of the elementary school is really to develop the mind and soul of the children and not merely to provide tame and acquiescent 'labour fodder.' If it is the national wish (as it is the wish of some individuals and groups) that the elementary school shall be simply an institution for the creation of subdued and 'patriotic' hands— institutions, therefore, with the minimum of material equipment and with a staff of inferior child-minders, the question of education does not arise and no discussion of method or purpose is needed. If, however, it is desired that the elementary school shall be an instrument of national education in the full sense, our two propositions must be considered. Upon the first we need not dwell very long. The safety of the world and the future of civilisation depend upon the character and intelligence of the multitude. Rulers of whatever name or rank may rise and fall; the ultimate power to make or mar the world will always be with the masses. As long as the elementary school is the chief means of humanising the masses, it is the most important school in the country, and no thought or care can be too great for it. One of the gravest social errors of the last half-century has been the deliberate depreciation of the real public-schools of England, the elementary schools. "Damned Board Schools," Mr Fisher's emphatic friend called them. Set up damned Board Schools, staff them with damned teachers, and you get a damnable population. Obviously you cannot have a humane education in damned schools. Obviously you will not find first-rate men and women willing to take service in damned schools. To spend an enormous sum every year on elementary education and then to depreciate elementary education is a mad waste of money. Either the expenditure or the depreciation must cease. The schools that educate the greatest number of people in greatest need of education are surely of the greatest importance.

Let us consider that as established and pass to our second proposition, that English is by far the most important subject in the elementary schools.

The past half-century of compulsory education has had its great successes, but (as we have already pointed out) it can hardly be described as successful. From the teacher's point of view, I think it has been a series of well-meant efforts to do the wrong thing in the right way. It is the system that is wrong. It works in the wrong direction and for a wrong purpose; and that it has produced good results is a tribute to the wonderful skill of the teachers and the wonderful spirit of the population. Our elementary school system is Ptolemaic. It has chosen a wrong centre. It has always been based, either overtly or tacitly, upon the inculcation of the kind of knowledge that has a producible result or 'answer,' something that can be tested, examined, marked. You deliver over to five hundred children certain assorted packages of arithmetic or geography or spelling, and then after a lapse of time you ask for them again. If four hundred and fifty give them back, you win; if more than fifty have lost their packages you lose. That was the prescribed official system of elementary education for so many years that its evil spirit still terrorises us, although its body is dead. Calling that education is like examining churches for their percentage of public conversions and calling that religion; it is like compelling preachers to be concerned chiefly with the visible thing—with trying to fill the penitent forms instead of trying to fill the kingdom of Heaven. For fifty years teachers have been trying to make the elementary school-boy *know* something, when they should have been trying to make him *be* something. They have been trying to make him, not a man, but an epitome of information.

The prescribed curriculum of the elementary schools almost appears to have been based on a university pass course. The pass man is expected to know a little about several subjects. Diminish the requirements in each, make a few substitutions, and you have the elementary school curriculum—a very little about a great many subjects. Within the elementary system itself this diminishing to scale was elaborately pursued. Each standard of the school was the exact reproduction of every other on a slightly different scale The boy of seven in Standard I had precisely the same *shape* of work as the boy of thirteen in Standard VII, all that differed was the quantity. Turn to the Code and Schedules for the 'eighties and observe the pyramidally symmetrical tables in which all knowledge is arranged from the first standard to the seventh. Here, for instance, is the official course in geography as it stood in 1887:

STANDARD I (i.e. age about 7):
To explain a plan of the school and playgrounds. The four cardinal points. The meaning and use of a map.

STANDARD II (i.e age about 8):
The size and shape of the world. Geographical terms simply explained and illustrated by reference to the map of England. Physical geography of hills and rivers.

STANDARD III (i.e. age about 9):
Physical and political geography of England, with special knowledge of the district in which the school is situated.

STANDARD IV (i.e. age about 10):
Physical and political geography of the British Isles, British North America and Australasia, with knowledge of their productions.

STANDARD V (i.e. age about 11):
Geography of Europe, physical and political. Latitude and Longitude. Day and night. The seasons

STANDARD VI (i e. age about 12)·
Geography of the world generally, and especially of the British Colonies and Dependencies. Interchange of productions. Circumstances which determine climate.

STANDARD VII (i e. age about 13):
The Ocean. Currents and tides. General arrangement of the planetary system. The phases of the moon. (*Note:* In Standards V, VI and VII, maps and diagrams may be required to illustrate the answers given.)

Thus it will be seen that, after thirteen, there is no geography left for a child to learn; which is not surprising, since he has begun to wrestle with the meaning and use of a map at seven, and the physical geography of hills and rivers at eight.

Let us now turn to the official requirements in English for the same year. They represent, remember, *the total amount of instruction the elementary school child ever received in his own language and literature.*

STANDARD I:
Reading. To read a short paragraph from a book not confined to words of one syllable.

Writing. Copy in manuscript characters a line of print, and write from dictation not more than ten easy words, commencing with capital letters. Copy books (large or half text hand) to be shown.

English. To repeat twenty lines of simple verse.

STANDARD II:

Reading. To read a short paragraph from an elementary reading book.

Writing. A passage of not more than six lines from the same book, slowly read once, and then dictated word by word. Copy books (large and half text hand) to be shown.

English. To repeat forty lines of poetry and to know their meaning. To point out nouns and verbs.

STANDARD III:

Reading To read a passage from a more advanced reading book, or from stories from English history.

Writing. Six lines from one of the reading books of the Standard, slowly read once and then dictated. Copy books (capitals and figures, large and small hand) to be shown.

English. To recite with intelligence and expression 60 lines of poetry, and to know their meaning. To point out nouns, verbs, adjectives, adverbs and personal pronouns, and to form simple sentences containing them.

STANDARD IV:

Reading. To read a few lines from a reading book or from a History of England.

Writing. Eight lines of poetry or prose, slowly read once, and then dictated. Copy books to be shown.

English. To recite 80 lines of poetry, and to explain the words and allusions. To parse easy sentences, and to show by examples the use of each of the parts of speech.

STANDARD V:

Reading. To read a passage from some standard author, or from a History of England.

Writing Writing from memory the substance of a short story read out twice; spelling, hand-writing and correct expression to be considered. Copy books to be shown.

English To recite 100 lines from some standard poet, and to explain the words and allusions. To parse and analyse simple sentences, and to know the method of forming English nouns, adjectives and verbs from each other.

STANDARD VI:

Reading. To read a passage from one of Shakespeare's historical plays, or from some other standard author, or from a History of England.

Writing. A short theme or letter on an easy subject: spelling, hand-writing, and composition to be considered. Copy books to be shown.

English. To recite 150 lines from Shakespeare or Milton, or some other standard author, and to explain the words and allusions. To parse and analyse a short complex sentence, and to know the meaning and use of Latin prefixes in the formation of English words.

STANDARD VII·

Reading. To read a passage from Shakespeare or Milton, or from some other standard author, or from a History of England.

Writing. A theme or letter. Composition, spelling and hand-writing to be considered. Note books and exercise books to be shown.

English. To recite 150 lines from Shakespeare or Milton, or some other standard author, and to explain the words and allusions. To analyse sentences, and to know prefixes and terminations generally.

Is it not pitiful? But why, it may be asked, should we turn to the obsolete schedules of the 'eighties? *Because those obsolete schedules still represent the spirit in which elementary education is understood by most officials and teachers.* The elementary school is still moulded after that pattern, the only difference being that it is moulded by the choice of the teacher instead of by the compulsion of the code. The school syllabus of the present is just as absurdly pyramidal—a little of everything in Standard I, a little more of everything in Standard II and so on The ideal 'time-table' of to-day is a mechanically symmetrical document recording the proud fact that every child in every class is taking the same subject at the same time That is the kind of time-table that most teachers and inspectors love in their hearts, whatever they may profess with their lips. The dead hand of the old code, embodied in the clutch of the compulsory time-table, is still heavy on the schools. A time-table for an elementary school is not valid until it has been approved and signed by an inspector, and any deviation from it must be recorded in the school log-book. What a 'subject' may mean can be gathered from the fact that the Time-Table Form supplied by the London County Council to its elementary schools has a printed summary on which is to be shown the time allotted to *each* of the following 'subjects': (*a*) Composition, Written; (*b*) Composition, Oral; (*c*) Dictation; (*d*) Grammar; (*e*) Reading; (*f*) Recitation, (*g*) Word-building; (*h*) Hand-writing; (*i*) Literature Teachers have been known to add even to that amazing list.

Instruction of this kind given in this spirit is not education; and, so far from being 'practical,' it has no reference to any real needs of the children. Come into a London elementary school and see what it is that the children need most. You will notice, first of all, that, in the human sense, our boys and girls are almost inarticulate. They can make noises, but they cannotspeak. Linger in the playground and listen to the talk and shouts of the boys; listen to the girls screaming at their play—listen especially to them as they 'play at schools'; you can barely recognise your native language. Go into a class-room, and ask each boy his name and address. You will hardly understand the familiar names, you will understand none of the unfamiliar names; you will barely recognise the numbers, and, if the district is strange to you, you will probably understand none of the street names[1]. Ask a boy to tell you something—anything, about a book, or a game, or a place, and he will struggle convulsively among words like a fly in a jam-dish. Ask him to write, and he will produce something like this, which has chanced to reach me on the very day when these lines are being written:

OxFord Ward
St. Georges Hospital
Hyde Palk Corner

Dear Sir

When we Broke up thurs I went up the Palk we Played a game of Football, and we had Three Mins to go to Finish. Well the out side Left of their team kick the Ball to me it happened to go a goal kick But the misfortune came then for he could not stop himself From running he knock me Heads over Heels in the Coal and I thing it twisted me angle, wel me Friends carried me to the antic festic Hospital in the Battersea Bridge Road. They told they could nothing has the Dotor had gone so I was Brought to st Georges. I was put under gas and what they did I don't know But when I woke I was sick. Well I must Closed this Letter now

I remain
yours affectionate
A ..... L.........

Excuse Writing I can't write much sitting in Bed
Goddbye
P.S. I shall be glad when I get back to School.

[1] I thought I knew my own school district very well, but I was puzzled the other day by a boy of eleven who told me he lived in 'Cow Street' That, at least, was all I could make of his clearest repetitions, and I had to ask him to write it down, when I found that what he was trying to say was 'Cale Street'

This letter, which has its attractive and appealing qualities (I feel rather uncomfortable about quoting it), was written by a boy of 13 years 9 months, in Standard VI—that is, almost at the end of his day school course. That the boy should write to his school at all is an intimation of a sentiment and attachment of which far too little use is made, especially by those responsible for the provision and maintenance of schools. The loyalty and affection of a mere elementary boy does not count for anything in the school statistics, and is therefore of no importance to the official mind. The public school boy has all the gracious amenities of hall and cloister and playing-field inviting and cherishing his loyalty. What gracious amenities do we offer the public elementary school boy to make his school a hallowed memory? We urge him severely to become even as the public school boy; we bully him in the press because he hasn't the Eton spirit; and, often, we don't give him so much as a hall in which it is possible for his school to assemble in a regular act of fellowship.

The letter I have quoted is no worse than you will find in any Standard VI elsewhere. You will notice the significant fact that some of the harder words are right and many of the easier words are wrong The boy can manage quite well a mechanically difficult word like 'misfortune,' which he will seldom need to use, but he does not know the crucial difference between 'as' and 'has,' or between 'kick' and 'kicked,' which he will constantly need to use. That might be taken as an allegory of the whole elementary system. We search out and overcome the difficulties that matter very little, and ignore the significant difficulties that stare us in the face. Our bright seventh standard boys know how to use a table of logarithms and don't know how to behave at table. We teach them with great labour to babble of isotherms and isobars, and send them out of school unable to speak their own language. Their reading is no better than their writing. In a popular story, Mr W. J. Locke thus dismisses (with some heat) one of his minor characters:

He hated travel and all its discomforts, knew no word of a foreign language, knew no scrap of history, had no sense of beauty, was utterly ignorant, as every single one of our expensively State-educated English lower classes is, of everything that matters on God's earth.

I do not wish to magnify the importance of Mr Locke; but his outburst is evidence of a kind, and some of it, at least, is true. As a humanising agent the elementary school has in great measure failed, in spite of much devoted and uncovenanted work by teachers and members of public bodies. I am dealing here with the only schools for which I have any right to speak; but I feel bound to add that the secondary schools and the public schools, with far greater opportunities, have also failed to create an enlightened, humanely educated population. We are still a nation of Barbarians, Materialists and Philistines. The only change since Matthew Arnold's time is that the classes have become a little more mixed. Mr Locke's impassioned reference to the costliness of our system is quite justified. "State education of the lower classes" is indeed a grotesquely prodigal dissipation of public money—as wasteful as if the state launched out into building without knowing whether it intended to build a workhouse or a factory or a prison or a palace, and stopped building at the very moment when the erection began to look like anything at all. The provision of part-time education in a Day Continuation School does not materially alter the circumstances, and I do not think such a system will ever be successful. The attitude of a boy at work cannot possibly be the attitude of a boy at school. The crucial fact is that the poor boy goes to work just when the wealthier boy goes to Eton; and we affect to be surprised at the difference of result. There is, as every man of business knows, a point at which a refusal to venture a little more money is a fatal economy. It is just at this point that we refuse to go on with the general education of our children. We drag our boys and girls away from school just when the period of creative development is beginning, thrust them into factories, mines, shops or offices, and then (like Mr Locke) curse them for being uneducated. For they *are* uneducated—we have only to look around us to discover that. And to offer them a continuation of their education is a mockery. In a very real sense their education has never been begun! What they lack most of all is language, the means of all intellectual activity and humane development, and it is language about which we seem to care least.

The tragic position in the elementary school is that *English cannot wait*. Other subjects can, and yet it is to them that our chief efforts are directed. If a boy is not taught arithmetic or geography or science, he will simply lack arithmetic or geography or science for a time; but if he is not taught good English he will be perfecting himself in bad English. Indeed, for elementary school children and for many secondary school children, the position is worse than that. Even while the schools may be teaching good English, the surroundings of street and home will be teaching bad English The teacher of arithmetic in the school has not to fight against the assembled powers of bad arithmetic outside; but the teacher of English is continuously assailed by powerful and almost insuperable hostile forces He must expect to gain, not an absolute advantage over simple ignorance, but a bare balance of advantage over elaborate opposition. The teacher's hardest struggle is not against pure ignorance but against evil knowledge. So hopeless does the struggle seem that many elementary school teachers give it up, and say that the attempt to teach good English to children who live and move in an atmosphere of degraded English is sheer waste of time. I taught for twelve years in Rotherhithe and Poplar, and I sympathise with their feelings; but I believe their view is wrong. I believe that the more powerful the forces of evil the more powerful and unremitting must be the opposing efforts The Harrow master may, perhaps, relax his vigilance for English, the Hoxton master never. A poor, intelligent boy who is compelled to come to school has a clear right to have his language cleansed and purified, and we must accept the burden of effort. The teacher's business is not simply to lay bricks on an empty foundation, he has first to clear the cumbered ground, and begin his edifice, as he clears, with what appears to be a heap of rubbish. In plain words and in the ordinary sense, English is not a school 'subject' at all. It is a condition of school life.

The difference is important, for the failure of the elementary school to be a humanising influence on its products may be attributed in part to the mistaken efforts of the teachers in treating English as they treat Arithmetic, namely, as a mere subject, with a limited matter of its own, a fixed place in the time-table and a right to no more than a share in their daily exertions. Those exertions have been in the past not only limited, but sometimes misapplied, with the result that, of all school lessons, the English has usually made least effect upon the pupils' minds and the least appeal to their liking. Even to-day, when, as I have admitted, methods of teaching have been reformed and the matter taught has changed beyond recognition, there is evidence that teachers have not really envisaged the right dimensions of English in the schools, and that they are still expending their very admirable enthusiasm and amazing skill upon it as a specific and limited subject, or, worse, as a collection of detached subjects. But, as we have said, English is really not a subject at all It is a condition of existence rather than a subject of instruction. It is an inescapable circumstance of life, and concerns every English-speaking person from the cradle to the grave. The lesson in English is not merely one occasion for the inculcation of knowledge; it is part of the child's initiation into the life of man.

Another peculiarity of English that gives it special importance is that it is the medium of instruction in school—as, indeed, it is the medium of all intercourse, social and commercial, public and private. What should have been its strength in school has been its weakness. As the medium of instruction it has been everybody's business, and has, therefore, become nobody's business. Teachers seem to think that it is always some other person's work to look after English. *But every teacher is a teacher of English because every teacher is a teacher in English.* That sentence should be written in letters of gold over every school doorway. Teachers are very specially the official guardians of the English language. We cannot give a lesson in any subject without helping or neglecting the English of our pupils. One of the most useful lessons in economy and lucidity of speech I have seen was actually a practical geometry lesson, in which the teacher required boys in turn to come out and give the class exact directions, step by step, for the working of certain problems. Ability of the pupil to make a concise and lucid statement is postulated in our teaching of every subject; but how many of us ever try systematically to cultivate it? Yet without clearness of expression clearness of thought is impossible. No one can set down clearly what is not clear to him; and the effort to secure clearness of expression is a great step towards clearness of thought. The thought may be wrong; but the very clearness of a difficulty helps us to clear it up. English reacts everywhere. Teachers of science complain that their work in science is wasted, not because their pupils cannot make observations, but because their pupils cannot record observations. Teachers of languages complain that their pupils are confused by a foreign language because they are confused by their own. That is, progress in science, or French, or German, is impeded by faulty English. But let this be clear; no teachers, whether of science, or languages, or mathematics, or history, or geography, must be allowed to evade their own heavy responsibilities. They must not say "Our business is to teach Science or Mathematics or French, not English." That is the great fallacy of 'subject' teaching. It is very definitely their business to teach English, and their failure to recognise it as their business is a cause of the evil they deplore. In a sense the function of history, geography, science and so forth in school is to provide material for the teaching of English. The specialist teacher defeats his own purpose precisely to the extent to which he neglects the language of his pupils.

But in no sort of school is the English more sacrificed than in the elementary school, where it is needed most. Elementary education has failed—has failed nobly—because it has always attempted too much. It has tried to accomplish something when it should merely have tried to begin something. The elementary school is nothing but a preparatory school and it has tried to be a university. Knowing how many thousands of children leave school in their earliest teens never to go to school again, the teachers have striven (and have been driven) to cram into their pupils during five or six years the information that more fortunate boys get in the maturity and leisure of their public school and university period. Consider the wickedness of trying to turn out of the elementary school children of barely fourteen (from poor and bookless surroundings), equipped with a clear view of English dynastic and political history from Caractacus to George V (not forgetting the history of the British Empire, the Near East and the United States of America), with General Elementary Science (or Nature Study) and the geography of the Five Continents and the Seven Seas thrown in; and that in addition to facility in elaborate arithmetical calculations (some of which are never used in these days of tables and slide rules), and a command of English composition from paragraph to précis and of spelling from Abacus to Zymotic—all that, and more, at fourteen, when it is obvious to every intelligent person that what the elementary school children need most sorely is to be made humanly articulate and aware of the decencies and amenities of life! English is thus not merely something for them to learn, but the condition of their learning everything else. All their immediate and future intellectual progress, all their developing emotional powers, all their social and industrial existence must be built, if they are to be built successfully and enduringly, upon the foundation of language communicated and communicable. By language they will learn almost everything; without language they will learn almost nothing. It is the special duty of the elementary school to lay that foundation of language and leave later years to determine the nature of the building[1].

A third peculiarity of English is that, for the immensely greater number of English-speaking people, it is the only language ever learnt, and thus the sole means of linguistic study, the sole means of approach to literature. The word 'sole' is used numerically and not in disparagement. Unfortunately it has been the habit of many English leaders of thought to speak of their own language with at least an implication of disparagement. Most of these leading men belong to the public school and university

[1] I am not anxious to drag that over-rated person 'the practical business man' into a discussion of this sort, but I happened, at the moment of reconsidering this page, to hear something very apposite from the staff manager of a world-famous business concern. He knew nothing of any special interests and predilections of mine, nothing beyond the fact that I was a teacher. I will try to reproduce his words "If I may presume to offer my views about education in the Council Schools, there are two things that I think are more important than anything else. The first is the English language. Every boy ought to have a thorough grounding in his own language. Anyone can put words together to make some sort of meaning, but it takes a lot of work to make a boy write a simple, clear and correct letter. The next important thing is speech. Many a boy's prospects here [in the particular firm] have been spoiled because he can't speak correctly and because he has a Cockney accent You ought to have eradicated the Cockney accent by this time! You gentlemen in the schools seem to spend all your time in teaching the boys things they ought to be learning three or four years later. What is the result? They leave school with a lot of information that they forget in a few months, and without any of the grounding in their own language that would stay by them all their lives " It is only fair to say that this manager was speaking with the requirements of his own firm in view, but even so, my case could scarcely have been put more strongly.

tradition, a tradition endeared to them by the sanctities of place and the joys of remembered youth. They were educated in a certain way, and it seems to them impossible that there can be any other way; and so their affectionate praise of a classical education always comes with an implication or a declaration that English is necessarily an inferior instrument. I suppose it is literally true that some famous English scholars never had an English lesson in their lives. Their English was always a by-product of home and school and university. Happy the few who get it thus; but they are the few, not the many!

I cannot speak with authority about the kind of education I did not receive; but as a teacher I feel competent to criticise the arguments usually advanced by the extreme classical right. Perhaps the most striking fact in a long controversy is that, although a classical education is claimed as a unique instrument of humane culture, no one has demanded that it should be applied to the lower and lower middle class population, which is plainly in need of some improving treatment. Why? If its value is so great, it ought to be given the widest possible application in a nation of freemen with political rights and power. Surely the weak point about 'compulsory Greek' is that it never was compulsory. A compulsion that touches only a tiny fraction of the population is a compulsion *pour rire*, and no compulsion at all. If Greek is so precious that people must be made to learn it for their good, then, obviously, all people must be made to learn it. If we leave without protest the great population entirely un-affected by something alleged to be necessary, we stultify all our arguments for the necessity. In India, we are told, the despised *sudra* must not listen to the sacred books and must not memorise any passages from them. These privileges are restricted to persons of caste. Can it be that the unrecognised or unproclaimed motive of our classical extremists is the perpetuation of social inequalities —one kind of education for the poor and another for the wealthier? That object may be very desirable, but then it ought to be made unmistakably clear as an object. If it is understood that a classical education is merely a class education writ large, much recurrent controversy will be prevented. But, apart from all that, the large claims made for the effect of a classical education on the mind cannot be established. Some of these claims are based on the now discredited belief in the possibility of a general 'mental gymnastic' or formal training obtainable from one special subject. The palpable failures of the classical system poured out from the public schools surely disprove any high claims for the peculiar efficacy of that discipline. Whatever is trained in the average agreeable products of the public school it is certainly not the mind. But there have been some shining successes, and it is to these that our attention is always directed. They certainly deserve our attention; for it soon becomes obvious to any careful observer that they are the kind of men who would have been successful products of almost any system, even of almost no system. The virtues they have modestly attributed to their classical education are really the virtues of their own minds—minds that have taken a wonderful polish under classics, and would have taken a wonderful polish under anything. But, even if we admit (as we may) that they probably benefited more by classical study than they would have by any other, the important fact for us to notice is that their education was in the main *a literary education*—an education in humane letters. That the vehicle of that education was Latin and Greek is important, but not of the greatest importance. The cause of the classical languages is always assumed to be the cause of the humanities; and the cause of the humanities is nearly always assumed to be the cause of the classical languages. Whenever there is an outbreak of that perennial battle of the ancient and modern curricula, the classical combatants will almost invariably be found to be defending one thing under the impression that they are defending another. That is to say, the eloquent champion of compulsory Greek is often not defending Greek at all, but urging the claim of humane studies. And one very unfortunate result of this is that those who are moved to attack the study of what they call dead languages have the appearance of attacking the humanities. Sometimes, indeed, but less often than people suppose, they actually are attacking the humanities in order to press the claims of physical science, and of science we may say briefly that though for the few it may be as humanising as the classics, for the education of the many it is even more arid and unprofitable. The cause of the classical languages is certainly not the cause of the humanities. The two overlap, but they are not coincident; and it is highly important that we should recognise

this. The possibility of a general humane education lies precisely
in the fact that the humanities are far wider than the classics.
For consider what a good classical education really means. It
means, as a first step, the gradual acquisition of a language, its
vocabulary and its simpler mechanism; then, later, a more ex-
tended acquaintance with vocabulary and mechanism, and a
struggle to achieve a higher standard of correctness; and then,
later still, an habitual association with the beauty, delicacy and
refinements of human speech; that is, with the possibilities of a
marvellously delicate yet elaborate organism having a life—a sort
of living logic—of its own. It means an acquaintance with law and
order, with sanctions and implicit prohibitions, a submission to
grace and strength and economy and power, and the recognition
of a force that can disregard old laws and make a law of its own—
i.e. with development and growth. And together with all this
such an education means a gradual, leisured and comprehending
acquaintance with that crystallisation of personality, life and experience which we call great literature, and with the history,
science and philosophy that, in natural consequence, form
part of the completeness of literature. Such a progressive,
co-operative initiation into the uttered and embodied life of man
we call a humane education; but it is certainly not the prero-
gative of two tongues alone of all the world's speech to give it.
For some, for a few, the Greek and Latin languages are the gates
that open immediately into Paradise; for others, for the many,
they are gates, shut and barred, at which they knock and wait in
vain when they might have found an open door elsewhere. We
can cheerfully concede to the classical enthusiasts almost as much
as they claim; but we must never concede to them that Greek
and Latin are the only means of access to the gardens of the soul,
and that, these unopened, nothing remains to mankind but the
bleaker plains of text-book knowledge. We need not spend any
argument upon those who, in trying to imitate the raptures of
their betters, have adopted a cant of the classics as a public intimation of their social superiority. They have their reward. Nor
need we take much notice of those who (as I have heard with my
own ears) adduce the classics as the origin or source of English
literature, and maintain that, as Sophocles wrote plays two
thousand years before Shakespeare, it is necessary to read Sophocles

.

before it is possible to understand Shakespeare. I am reminded of the early nineties when the later works of Wagner were still esoteric, and were forcing their way with difficulty upon the London stage. I was given to understand by the elect that before I could attain to any appreciation of the *Ring des Nibelungen* I must study the principles of Greek drama, the economics of Marx and Lassalle, the politics of Bakunin, the doctrines of Buddha, Schopenhauer and Feuerbach, the history of opera, the compositions of Gluck, the interminable prose works of Wagner himself, the history of Teutonic mythology, the difference between an opera and a music-drama, and the dramatic significance of the musical themes or motives in the whole work. All this being accomplished, I might then be fit to go to a performance. Really, it is very like the contention that a boy must labour in youth at the *Aeneid* in order to read *Paradise Lost* when he is old, at Theocritus in order to read *Lycidas*, and at Plato in order to read Shelley. It is all very strange. The earnest classical master who would pronounce Milton and Browning too difficult for the boy of twelve has no hesitation in making him read Virgil and Horace. I found in music that the best way to an appreciation of any composer was to read the scores and hear them played as often as possible; and I know that in poetry the way to approach an appreciation of Milton or Shelley or Wordsworth is simply to read Milton and Shelley and Wordsworth. Indeed, it may even be better to read first our contemporary Masefields and Hardys and work backwards to the older masters—at least, Charles Sorley found it so. When we have read Shelley we can find a special interest in Plato; but we shall not get at the beauty of Shelley by wrestling first with the language of Plato. Neither in art nor in science can we begin at some arbitrary point called the beginning: we have to begin at a very clear point called the end—our end of knowledge, not the other undiscoverable end.

What guidance in English can we get from this old controversy about the classics? The first point to note is this: that for every one of the choice spirits who get their splendid culture from the classics, there are hundreds of others, their social equals, who get very little culture at all, because they can never break through the barriers of a foreign language. The second point is this: that those barred out and forbidden hundreds can receive an excellent

education in and through their own language. The British naval
officer is not conspicuously inferior to the products of the Army
Class, and he is educated at a school where the classics are not
taught at all. The third point that emerges is this: that even if
we granted all the claims made on behalf of a classical education,
the plain fact stands that the greater number of English boys
and girls do not and cannot receive that education. A classical
education is something for the aristocracy of mind, in whatever
social class it is found, it is not an education for the masses, in
whatever social class they are found. Let us hear no more, then,
of the parrot cry that no one can be disciplined in humane letters
save through the medium of the classical languages. The assertion
is untrue, and an impediment to national education. We don't
want *Anathema* in the school room. Let *Quicunque vult* and
threatened excommunications of all kinds remain in the realm
of theology and outside the realm of education. After all, it
should need no heroic self-denial in anyone to refrain from asserting
that the language of Shakespeare and Milton and Shelley, of Bacon
and Burke and Gibbon, of Sidney and Spenser and Ralegh, of
Lamb and Swift and Johnson, is mysteriously disabled from con-
tributing to the making of an Englishman.

Thomas Huxley was not only a great man of science, but an
excellent writer of English. These words of his, therefore, deserve
respectful consideration:

Every Englishman has, in his native tongue, an almost perfect instrument
of literary expression; and, in his own literature, models of every kind
of literary excellence If an Englishman cannot get literary culture out
of his Bible, his Shakespeare, his Milton, neither, in my belief, will the
profoundest study of Homer and Sophocles, Virgil and Horace, give it
to him.

This discussion of the classics is neither irrelevant nor unnecessary, because, unfortunately, among the hostile forces to be
encountered by those who believe in the value of English for the
English must be reckoned those who persistently urge the claims
of Greek and Latin, and consider English no language for a
gentleman. It seems to be held that English boys cannot be
taught English, and, on the whole, need not be taught English.
What is necessary is to teach them Greek and Latin, and then
in some mysterious way, English will come. There are possibly

some old public school men whose instincts of loyalty will urge them to agreement. But is there, I ask, any other country in the world, where men of intellectual standing will be found assuring their countrymen that carefully selected boys of the most favoured class in the land cannot learn their own language without the special support of two ancient tongues? Statements of this sort are merely the *clichés* of tradition. They are delivered with the authority of the centre, but they sound remarkably like provincialism.

It may be said, I hope without offence, that the classics are exotics, and I think we must beware of cultivating the exotic habit—the frame of mind, that is, in which we find nothing good in what is around us, but must scour the seas and dislocate the seasons to gratify ourselves. There is a story, apparently frivolous, that can be read as a parable A certain epicure, who loved all things out of season, who ate strawberries in December and scorned them in June, who found green peas delicious in February and uninteresting in July, once asked some friends to a dinner in May at which he promised them a very great rarity. The expected dish proved to be new potatoes. When it was objected to him that new potatoes in May can scarcely be called a rarity, he replied very earnestly, " Ah! you don't understand, these are *next year's* new potatoes." Now, isn't that precisely what the classical extremists are always doing, looking in May for next year's new potatoes?

The classics in education must be described as a powerful vested interest, liable to show resentment if a rival claimant to a share in the humanities is brought forward. The classics can still claim the power of the purse and the pride of place, and can be vigorous in obstruction. No person who knows and loves our national inheritance of literature can regard the classical languages and their great tradition in our oldest schools with anything but reverence and affection. but that emotion is not incompatible with an equally sincere belief that, for English people, the great and immediate means of a humane education is to be found in English, and in no alien tongue whatsoever, either ancient or modern

It is as well to remind ourselves that the claim of English for the English has quite a respectable tradition Richard Mulcaster, the first Headmaster of Merchant Taylors' School (where he had Edmund Spenser for pupil) and later High Master of St Paul's, author of two most excellent treatises on education—*Positions* (1581) and *First Part of the Elementarie* (1582)—thus urges the peculiar value of English for the English as against all other tongues—I quote Mr Oliphant's modernized paraphrase:

It is the opinion of some that we should not treat any philosophical subject, or any ordinary subject in a philosophical manner, in the English tongue, because the unlearned find it too difficult to understand in any case, and the learned, holding it in little esteem, get no pleasure from it. In regard both to writing in English generally, and my own writing in particular, I have this to say: no one language is finer than any other naturally, but each becomes cultivated by the efforts of the speaker who, using such opportunities as are afforded by the kind of government under which he lives, endeavours to garnish it with eloquence, and enrich it with learning. Such a tongue, elegant in form and learned in matter, while it keeps within its natural soil, not only serves its immediate purpose with just admiration, but in foreigners who become acquainted with it, it kindles a great desire to have their own language resemble it. Thus it came to pass that the people of Athens beautified their speech in the practice of pleading, and enriched it with all kinds of knowledge, bred both within Greece and outside of it. Thus it came to pass that the people of Rome, having formed their practice in imitation of the Athenian, became enamoured with the eloquence of those from whom they were borrowing, and translated their learning also. However, there was not nearly the same amount of learning in the Latin tongue during the time of the Romans as there is at this day by the industry of students throughout the whole of Europe, who use Latin as a common means of expression, both in original works and in translations. Roman authority first planted Latin among us here, by force of their conquest, and its use in matters of learning causes it to continue. Therefore the so-called Latin tongues have their own peoples to thank, both for their own cultivation at home and for the favour they enjoy abroad. So it falls out that, as we are profited by means of these tongues, we should pay them honour, and yet not without cherishing our own, in regard both to cases where the usage is best and to those where it is open to improvement. For did not these tongues use even the same means to cultivate themselves before they proved so beautiful? Did the people shrink from putting into their own language the ideas they borrowed from foreign sources? If they had done so, we should never have had the works we so greatly admire.

There are two chief reasons which keep Latin, and to some extent other learned tongues, in high consideration among us—the knowledge which is registered in them, and their use as a means of communication, by the learned class throughout Europe While these two benefits are retained, if there is anything else that can be done with our own tongue, either in beautifying it, or in turning it to practical account, we cannot but take advantage of it, even though Latin should thus be displaced, as it displaced others, bequeathing its learning to us. For is it not indeed a marvellous bondage, to become servants to one tongue for the sake of learning, during the greater part of our time, when we can have the very same treasures in our own language, which forms the joyful title to our own liberty, as the Latin reminds us of our thralldom? I love Rome, but I love London better: I favour Italy, but I favour England more: I honour the Latin tongue, but I worship the English. I wish everything were in our own tongue which the learned tongues gained from others, nor do I wrong them in treating them as they did their predecessors, teaching us by their example how boldly we may venture, notwithstanding the opinion of some among us, who desire rather to please themselves with a foreign language that they know, than to profit their country in their own language, which they ought to know It is no argument to say: Will you dishonour those tongues which have honoured you, and without which you could never have enjoyed the learning of which you propose to rob them? For I honour them still, as much as any one, even in wishing my own tongue to be a partaker of their honour. For if I did not hold them in great admiration, because I know their value, I would not think it any honour for my own language to imitate their grace. I wish we had the stores with which they furnished themselves from foreign sources For the tongues that we study were not the first getters, though by learned labour they prove to be good keepers, and they are ready to discharge their trust, in handing on to others what was committed to them for a term.... The dishonour will lie with the tongue that refuses to receive the inheritance intended for it and duly offered to it, and from this dishonour I would our language were free. I admit the good fortune of those tongues that had so great a start over others that they are most welcome wherever they set foot, and are always admired for their rare excellence, disposing all men to think little of any form of speech that does not resemble them, and to rank even the best of these as marvellously behind them. The diligent labour of the learned men of ancient times so enriched their tongues that they proved very pliable, as I am assured our own will prove, if our learned fellow-countrymen will bestow their labour on it. And why, I pray you, should such labour not be bestowed on English, as well as on Latin or any other language? Will you say it is needless? Certainly that will not hold. If loss of time over tongues, while you are pilgrims to learning, is no injury, or lack of sound skill, while language distracts the mind from the sense, especially with the foolish and inexperienced, then there might be some ground for holding it needless. But since there was no need for the present loss of time in study through labouring with tongues, and since our understanding is more perfect in our natural speech, however well we may know the foreign language, methinks necessity itself calls for English, by which all that bravery may be had at home that makes us gaze so much at the fine stranger....

Our English is our own, and must be used by those to whom it belongs.

First Part of the Elementarie: Peroration.

Mulcaster died over three hundred years ago, and the obscure language of his little island has become the language of the world with a literature second to no other. But no attempt has ever yet been made to give the whole English people a humane education in and by the English language. What we need to do is to take the enthusiasm and enlightenment that marked the best classical teaching and transfer it to English. It was often said that classical masters were the best teachers of English. Obviously! for they were the only teachers who had received a literary education. At the Renaissance, when our speech seemed barbarous and our literature nothing but an old song, the English scholars who lit their torch from the lamp that Marsilio Ficino burned before the bust of Plato carried light into darkness, and, by giving us Greek, really gave us English They sang, and rightly sang, the glory of Greek, because it was a living tongue when English had not fully come to birth; but the pedant of to-day finds some special glory in Greek because it is dead. English has inherited; and those who still try to imitate the faded ecstasies of the Renaissance scholars are untrue to the Renaissance. The Renaissance was not so much a new birth as a resurrection. Why seek we, then, the living among the dead? Our monks still babble of Greek and Latin as the only way to literary salvation; and rather than abate a jot or tittle of that claim they are ready to let the millions of boys and girls from the elementary schools and the modern middle class schools, and the thousands of eager and energetic young men from the colleges of science and technology—all hopelessly lost to classics —pass into life untouched by the thoughts that breathe and words that burn in exhalations from the native soil We do not give our elementary school children even so little as the decencies of cultivated speech. It is simply the fact that, after fifty years of compulsory education, John Bull has far less facility in his own language than the despised foreigner in his. In whatever direction of knowledge the elementary school products may have progressed during that half-century, in English (and all that it implies) they have progressed least of all. We have called one novelist in evidence, let us call another. Thus writes George Gissing in *Thyrza*:

The working man does not read, in the strict sense of the word; fiction has little interest for him, and of poetry he has no comprehension whatever; your artizan of brains can study, but he cannot read.

That is unquestionably true—how true, even of the ambitious working man, let the teachers of W. E. A. and U. T. C. classes testify. The deficiency is just what is wrong with the class from which the clever artizan comes. The English workman woos knowledge for her dowry, not for her diviner charms—if ever he is moved to woo knowledge at all. At the back of his mind is the idea of 'improving his position in life,' when what most needs improvement is his posture towards life. The English lower middle class is an uneducated class, and therefore the special prey of the political humbug and the 'stunt' newspaper. The man who can only study is in a lower stage of development than the man who can read. He is one to whom the primrose on the river's brim is not even a primrose: it is *P. vulgaris*, and nothing more. He is the man who can never see the wood for the trees—the man who can never take a view of things, the man without vision, or sense of something afar.

It seems plain, then, that our Ptolemaic educational system is in need of reformation. Its centre of gravity must be changed from subjects that give the boy scraps of information to subjects that will make him a civilised articulate human being. So far, in the process of education we have been looking for power before capacity—we have been trying to force a spade into hands that can hardly wield a spoon. That the elementary school boy's education is cut short at the age when the public school boy's is beginning is a tragic fact that may be lamented, but must not be allowed to determine the range and character of the curriculum. The elementary school boy is going to have a vastly harder and more disagreeable life in manhood than the public school boy; but that is no reason why we should make his school days harder with tough masses of indigestible knowledge. Mark Twain is not usually regarded as an educationist; but upon education, as upon other human activities, he has something wise to say. His text is this maxim: "In the first place God made idiots. This was for practice Then he made School Boards." Here is the comment:

Suppose we applied no more ingenuity to the instruction of deaf and dumb and blind children than we sometimes apply in our American public (i.e. elementary) schools to the instruction of children who are in possession of all their faculties? The result would be, that the deaf and dumb and blind would acquire nothing. They would live and die as ignorant as bricks and stones The methods used in the asylums are rational. The teacher exactly measures the child's capacity, to begin with, and from thence onward the tasks imposed are nicely gauged to the gradual development of that capacity. the tasks keep pace with the steps of the child's progress, they don't jump miles and leagues ahead of it, by irrational caprice, and land in vacancy—according to the average public school plan. In the public school, apparently, they teach the child to spell cat, then ask it to calculate an eclipse; when it can read words of two syllables, they require it to explain the circulation of the blood; when it reaches the head of the infant class they bully it with conundrums that cover the domain of universal knowledge.

It is undeniably true that we do not fit instruction to the obvious needs and deficiencies of normal children as carefully as to those of mental and physical defectives. A sound educational system must be based upon the great means of human intercourse—human speech in spoken and written word. For English children it must recognise the great peculiarities of English already described: (1) that it is of all school activities the chief, the most important and the most practical, because it covers the whole life of man from the cradle to the grave; (2) that it is the one school subject in which we have to fight, not for a clear gain of knowledge, but for a precarious margin of advantage over powerful influences of evil; (3) that it is the medium of all instruction, communication, business and general intercourse, the basis of knowledge, the condition upon which progress depends; (4) that it is (for the majority) the only language learned, and thus the sole means of linguistic study and access to literature, the sole means by which the people can be placed in contact with the embodied feeling, thought and experience of mankind. That is to say, our chief task in school is twofold: (a) to treat English as a tool, and (b) to treat English as a means of access to formative life and beauty.

There is no class in the country that does not need a full
education in English Possibly a common basis of education might
do much to mitigate the class antagonism that is dangerously keen
at the moment and shows no signs of losing its edge. At present,
social antagonism is emphasised by the whole of our school system
—if anything so haphazard can be called a system The poor boy
in the elementary school receives a certain kind of education—
a kind, as we have seen, not at all adapted to his personal needs,
the wealthier boy, of precisely the same age and much the same
mental endowment, receives in the preparatory school, an entirely
different kind of education, equally ill adapted to his personal
needs. In the education of the poor boy and of the rich boy there
is nothing whatever in common except the profound uselessness
of that education. For some of the elementary children there
are various kinds of further education—central schools, day
continuation schools, junior technical schools, trade schools, and
I think even additional kinds are in contemplation. These schools
are fences and barriers meant to keep the elementary school
children from going into the schools of the 'respectable' classes.
A few actually reach the secondary schools as scholarship children
or 'free-placers'—and in some schools they are kept apart from
the 'paying' pupils. No elementary school children ever pass into
the public schools. Thus, our elementary schools, secondary
schools and public schools represent, not educational grades, but
social antagonisms. For instance, the late head-mistress of a well-
known girls' secondary school in London refused to hold any
communication with the head-mistresses of the elementary schools
from which she received the Junior County Scholars. It may be
quite right that social grades should be emphasised by the schools,
it may be quite right that there should be (as there is) a distinctive
'class' education; but we must not be astonished to find as its
inevitable product a distinctive class antagonism If we want
that class antagonism to be mitigated, we must abandon our system
of class education and find some form of education common to
the schools of all classes. A common school is, at present, quite
impracticable. We are not nearly ready yet to assimilate such a
revolutionary change. But though a common school is impracticable, a common basis of education is not. The one common
basis of a common culture is the common tongue.

III

A Programme

"He described himself as a senior English master in a London private school."
"Poor wretch," said I.
"That's what I thought, and the more he talked, the more I thought it."
H. G. Wells.

As a school subject English can be considered in six aspects—aspects of one thing, remember, not separate 'subjects,' sometimes not even separable subjects. I am writing with the elementary schools chiefly in my mind; but the course here suggested should form the 'elementary education' of all English boys and girls, no matter what their social class may be or what their schools may be called. Naturally the amount of stress laid upon each aspect of the subject will vary with the immediate needs of the pupils. Thus, a child whose ear has been accustomed to the pitch, intonation and accent that mark the speech of cultivated people will naturally need very little of the training that must necessarily be given to elementary school children. These are what I call the six aspects of the English course: (1) Training in Speech; (2) Training in Talk; (3) Training in Listening; (4) Training in Writing; (5) Training in Study; (6) The Induction to Literature. Let us consider each of these in detail.

I

Systematic training in standard English speech

This is a difficult and disputable part of the course. I give here my own view, but I know it is a view that many refuse to take. I believe, however, that, within certain limits, most teachers are in agreement.

This country is torn with dialects, some of which are, in the main, degradations. Enthusiastic 'localists' cling to their dialects

—and cling, sometimes, to the merely ignorant mispronunciations, blunders and lapses which they fondly imagine to be part of dialect. The untutored speech of the multitude does not necessarily represent the unspoiled freshness of a beautiful patois. Thus, to quote a simple example, the majority of English people—even some who pretend to education—talk of 'laying down' when they mean 'lying down.' Are we going to protect what is nothing but a piece of ignorance on the ground that it is part of the natural common speech? I feel, myself, that the schools have nothing to do with any patois. The language of all English schools should be standard English speech. It is not the business of teachers either to cherish or destroy a local dialect; they have simply to equip their pupils with the normal national speech—as a sort of second language, if the grip of patois is very strong. Most educated Englishmen who spent their youth in a dialect area are bi-lingual —indeed the desire to write in Doric is an eternal passion in the literary breast. Standard English need not be fatal to local idiom Where a dialect is genuinely rooted it will live; where it is feeble but curiously interesting, it may be kept artificially alive by enthusiasts; where it has no real reason for existing it will perish, as all provisional institutions must perish. Even if the school tends to extinguish a local idiosyncracy of speech, it is not necessarily doing evil. No serious damage is done to national tradition if a boy is taught to say 'I'll hit him' instead of 'Us'll hit he.' He will fight all the same. It seems clear to me that every English child, whether he has a local speech or not, should be taught in school the standard English of the day. There is no need to define standard English speech. We know what it is, and there's an end on't. We know standard English when we hear it just as we know a dog when we see it, without the aid of definitions. Or, to put it another way, we know what is *not* standard English, and that is a sufficiently practical guide. If any one wants a definite example of standard English we can tell him that it is the kind of English spoken by a simple unaffected young Englishman like the Prince of Wales, or by Mr Balfour (born in Scotland) or by Mr Asquith (born in Yorkshire), or by Mr Dennis Eadie or Mr Gerald du Maurier or Miss Ellen Terry or Miss Henrietta Watson—actors as a rule speak with less affectation than actresses We want English people to speak English as unaffectedly as

Coquelin spoke French. French people do not all speak alike (as the untrained English ear is apt to suppose), but there is much less divergence from the centre in France than in England. We have no need, however, to go to France for an example One of my most cherished memories of happy nights at the theatre in days of youth is the Buckingham of Forbes-Robertson in Irving's production of *Henry VIII*. Buckingham is not so much a character as a speech; and that speech as Forbes-Robertson delivered it, was a model, or rather a monument, of beautiful English utterance That sort of thing, we can say, is standard English speech at its noblest. We need waste no time upon the people who hold up hands of horror at the thought of everybody's speaking just like everybody else What we have to look at is the first possibility of reform, not the theoretical extremity of accomplishment. No form of living speech can be stationary even though a standard be fixed. Everybody will not speak like everybody else. There will always be some divergence, as there now is among people of undoubted cultivation, to say nothing of the personal overtones that we could not eliminate even if we tried. What we want to do, what we must do, is to lessen the extremes of difference that make Whitehall, Whitechapel and Whitehaven foreign to each other.

What we must not do is to set up affected suburban 'refinement' as a standard. One out-spoken young person in a factory continuation school, when speech training was mentioned, declared that she wasn't going to talk like 'the rich people.' She added scornfully, "They talk as if they weren't well. They pinch their mouths up and say, 'Oo, noo,' and 'Thenks offly.'" Well, her decision might have been strengthened by a very impressive lady who sat opposite to me in a 'bus the other day and asked quite unmistakably for a ticket to 'Sloon Squah,' thus meeting, as extremes will, the aspiring young shop-girls whom I hear every morning at Hammersmith asking for tickets to 'Natesbridge.' It is distressing to hear a London elementary school sing, in a well-known hymn, "Pryse Him for His gryce and fyver"; but it is just as distressing to hear a church choir in a highly respectable suburb intimating a similar gratitude for 'grease and fever.' A man's speech is usually his label for life, and we ought to see that good boys are not wrongly labelled.

As the 'O' revealed Giotto (writes the Autocrat)—as the one word 'moi' betrayed the Stratford-atte-Bowe-taught Anglais—so all a man's antecedents and possibilities are summed up in a single utterance which gives at once the gauge of his education and his mental organisation.

Possibilities, Sir? said the divinity student, can't a man who says *Haow?* arrive at distinction?

Sir, I replied, in a republic all things are possible But the man *with a future* has almost of necessity sense enough to see that any odious trick of speech or manners must be got rid of

The desiderated training in speech must begin in the earliest stages and continue without any lapse through the child's school life. At first the training will be purely imitative, but later it should mean instruction in the use of the speech-organs. Needless to say that a teacher of speech untrained in phonetics is as useless as a doctor untrained in anatomy. The teacher without a phonetic training is unable to make any systematic correction because, as a rule, the untrained ear cannot isolate the real cause of the trouble —often, in London speech, for instance, a defect of one element in an unrecognised diphthong. Speech-training reveals, at times, physical defects that need specific remedial or surgical treatment Defects of the speech-organs are often overlooked at medical inspections. It is unnecessary to do more than barely mention how immensely the training in English speech will benefit every child who comes later to learn a foreign language.

Allied with pronunciation and the production of sounds is clearness of articulation. This is a very important aspect of English teaching and it is frequently overlooked. Slovenly, lazy articulation must never be tolerated in any class of any school of any grade The affected 'refinement' of one social group is as abominable as the unaffected lack of refinement of another. Many mistakes in spelling are really mistakes in articulation. The London boy who writes 'he at to go' is not making a mistake in spelling; he is spelling correctly what he habitually says. The first correction must come in his speech. For most London boys there is a number called 'lem'; that they ever spell it 'eleven' is a mystery. On the explosive subject of phonetic or 'simplified' spelling I propose to say nothing at all. The teacher's obvious business is to teach the kind of spelling in public demand. As long as the general public demand is for the existing spelling teachers must teach it. Our present spelling is deplorable and makes the teaching of reading to very small children a terrible labour Still, both reading and spelling are taught; and though the child of seven spells with a wild and wonderful license, by fourteen he spells, on the whole, pretty well. More will be said about spelling in a subsequent section.

What seems to me a final answer to those who despair of teaching the decencies of speech to our elementary school children is the fact that these same children are taught to sing beautifully. Those who can be taught to sing can be taught to speak. But to accomplish this means hard work—it means unsleeping zeal and patience. The desired standard will not be attained in one generation or two, but every step will be something gained. Naturally much of the labour will fall, as it already does, upon the teachers in the infants' schools; and, in happier conditions, their work will be regularly continued in the next departments of the school, instead of being abandoned (as it usually is) for something more 'practical'—like arithmetic. I think specially of one infants' department, in which, owing to the personal enthusiasm of an enlightened head-mistress, speech training was a stressed activity, with the result that little fellows of seven and eight came up to the boys' department able to pronounce such crucial vowels for Londoners as the *a* and the *i* and the *ee* in *say* and *wild* and *wheel* quite prettily. But speech training had no place among the stern and useful activities of the boys' department, and in a very short time the little graces of speech had vanished, and *say* had become *sy, wild, wahl* and *wheel, wil,* as in any other London school. Now, let this sedulous training of speech become universal in the infants' school, and let it be continued as a compulsory and rationalised activity in the boys' and girls' departments, and what a difference we should make! In this country classes are sundered by difference in language—difference of speech is a symbol of class antagonism. How far a common speech would go to mitigate antagonism and remove misunderstanding—how far the removal of class difference is either desirable in itself or a proper aim of education—these are debatable questions, but surely if there is a unity called England there should be a unity called English!

And the matter goes further than the limits of this island.

A PROGRAMME

English is now incontestably the language of the world. W
should the standard of spoken English be found if not in Engl
But there is no standard here. Each county, almost each town, is
a law to itself and claims the right purity for itself. This is not
independence, it is merely provincialism; and it is not the duty
of the schools to encourage provincialism, but to set the standard
of speech for the Empire. The difficulties are not insuperable.
Individual efforts of particular schools are useless—there must
be concerted action. Let the central authority for education
clearly demand that in all English schools standard English speech
alone must be used, and we shall soon find our way in the right
direction without elaborate and disputable prescriptions A written
constitution in phonetics is not desirable. What does seem desirable
is that the teacher's own equipment in accent and intonation
should be more closely scrutinised than it is. I urge, then, as an
important and essential part of school activity the training of the
pupils in the sounds of standard English speech The English boy
has an indefeasible right to the King's English.

II

Systematic training in the use of correct spoken English

The person who cannot make a concise, correct, lucid and
intelligible statement in speech is an uneducated person, whether
he is a Fellow of All Souls or a fellow from the slums. The importance of this 'talk-training' is so great that little need be said
about it. It ought to be the most successful feature of school work,
simply because the interchange of views between teacher and
taught is the staple of most good lessons, but it is very obviously
not the most successful feature of school work, as any visitor will
at once discover. Too often teaching is little more than the delivery
of monologues by the teacher, the pupils' required contribution
being dead silence Even when the class is asked questions, the
teachers are usually content to accept the intention of an answer
(sometimes a very good thing to do)—that is, to accept answers
for the sake of facts and to regard the form of expression as unimportant or beyond their control This is not limited to the
elementary schools. But, in a very real sense, a fact badly expressed

is much less valuable both to speaker and hearer than the same fact clearly expressed—if, indeed, it can be said to be a fact at all before it has reached exact expression. We cannot here discuss the value of style in art—we must simply take it as indisputable that style is not an added and irrelevant adornment, but something that gives speed, force and direction to utterance. We shall not be so foolish as to demand from the speech of our pupils the grace that is called style; but we must not be so foolish as to think that in any aspect of life fact is everything and form nothing. The decencies and amenities of civilised existence amount to feeling embodied in form. In a very real sense it is form that makes the world go round.

Any lesson requiring the use of speech should always be a lesson in spoken English, whatever the subject of the lesson and the special domain of the teacher. The business of the science teacher is not to teach some impalpable abstraction called science, but science embodied in decent speech. The business of a person at table is not merely to eat, but to eat decently. He may eat enormously, rapidly, and even rapturously, but if he does not eat decently he eats badly; in short, he does not know how to eat. So, the boy who slobbers out history in smears and messes of words, simply does not know his history, even if the facts he has emitted are correct. The power to transmit knowledge is part of know- ledge. It is useless for a person to claim that his sentences mean something to him if they mean nothing to everybody else. Teachers must abandon utterly the tacit belief that it doesn't matter how a thing is said if the facts are all right. Hell is paved with the sentences that have failed to express their meaning. Teachers must also abandon their tacit belief that the geography teacher's business is to teach geography, not English. It is every teacher's business to teach English[1]. Remember that our chief work in

---

[1] Here is an actual example of what happens. In a certain continuation school, attended by pupils of a low intellectual order, the cookery note-books containing recipes dictated by the cookery-teacher were found to be full of incomprehensible rubbish—the girls having been unable to take down what was said. When this was pointed out to the teacher, she was very indignant and declared expressly that it was not her work to see that the girls could do dicta- tion. She further added that the girls had an hour's English lesson every week, and it was Miss So-and-so's business to teach dictation. She found general support and agreement in her repudiation of responsibility. Now the whole

school is to call out capacity. Give the boy a workmanlike ability to use his own language, and it doesn't matter much if he thinks that the Equator is also called the Meridian of Greenwich. If his English is put right he will be able to get his Equator right, but while his English is bad he will get nothing right.

Probably the least valuable lessons in spoken English are those specifically called Oral Composition lessons, the most valuable are the casual lessons that occur in every subject. It is much harder to be natural than artificial; it is much harder for a boy to acquire the regular habit of decent natural speech than to remember the formalities of an occasional harangue. This is not said to depreciate the value of oral composition lessons in the least, but to prevent a tendency of teachers to limit 'talk-training' to these lawful occasions.

Personally I should never put oral composition on a school time-table. To assign a limited definite time for practice in talk seems to me as absurd as to assign a limited definite time for practice in conduct. Informal training in talk should scarcely ever cease; and I should assume that when a teacher wanted a debate, or a set of speeches, or any other special form of utterance, he would use any composition lesson, or, indeed, any other lesson, if the matter were appropriate. 'Oral Composition' on the time-table is apt to imply that we need not concern ourselves with the children's speech at half-past ten on Monday because the oral composition lesson is at half-past three on Wednesday. Much of what is said in a later section about the teaching of composition can be applied to the teaching of conversation.

The correction of the pupil's spoken English is a difficult and delicate matter, calling for the utmost of a teacher's technical skill. Correction should be kindly and casual, intimated rather than impressed. If your children like you they will try to imitate you. The act of speech must not be made an ordeal of terror. I have never encountered any boys who resented correction genially made, and I have usually found that they really do want to learn how to speak nicely. Remember that when a boy is speaking to you very badly he is not trying to speak badly; he

of the present contention can be concentrated on that one point. It certainly is the business of the cookery-mistress to teach dictation—as long as she finds it necessary to use dictation.

is not even indifferent; he is trying to speak well, and failing. He is suffering from a disease of language which it is the teacher's business to cure. Remember too, that the best teaching is that which means friendly cooperation between teacher and taught. Friendly conversation is a mode of friendly cooperation.

There are few graces in which the Englishman is more deficient than the grace of easy speech. No one wants him to be perpetually talkative, but then, surely, no one wants him to be a sullen and silent boor. We want him to be able to talk, even though he prefers silence. Neither public school nor secondary school appears to do much for him in this respect, the elementary school appears to do next to nothing, and to be unaware of his need or its own neglect. In spite of its alleged preoccupation with the boy's ability to earn a living, the elementary school does not equip him with the chief tool of all trades and callings, clear and intelligible speech. And yet of all his deficiencies, this lack of language is the most obvious. At least that is so in London. The elementary school boy takes out of school at fourteen, unmitigated and unimproved, the debased idiom he brought into it at seven, and even in that he is but semi-articulate. Less time is spent in school on the spoken language than on any other activity, and yet none needs more—none so glaringly needs more. The elementary school fulfils neither any immediate or ultimate purpose of education as long as it is content to take the speech of the children as it finds it, and to leave it there. The most damaging criticism that can be uttered against the elementary school is to be heard in the speech of its products. Surely, the first healthy impulse of any kindly person confronted with a class of poor, inarticulate children should be to say, not, "I will teach you arithmetic and history and geography and science and drawing," but "I must teach you how to speak like human beings " What can literature possibly mean to children whose habitual misshapen and untaught speech bears no resemblance to what they see in print? For them, in a sense, English literature is in a foreign language. To speech the rest can be added. Correct and lucid speech is not only an ornament and grace of life: it is one of the first and last necessities of corporate existence There is no more 'practical' subject in the whole range of school work; there is no subject more generally neglected.

III

Regular practice in the art of listening

The six aspects under which we are considering English are not clearly defined and exclusive. Indeed, they cannot, they must not, be made exclusive. Thus, our first two aspects shade into one another, and much that we have said about speaking applies equally to writing. This third aspect, the art of hearing, cannot be dissociated from the others—it is a part of speech, of writing, and of reading. As, however, listening seems specially associated with talking, let us consider it here. Very little need be said to indicate its importance in education. A poor teacher or lecturer can make an audience endure; a good teacher or lecturer can make an audience listen. Unfortunately everybody is not interesting, and listening involves much disciplined attention. How far do we give our pupils practice in listening? There is a passage in Mrs Piozzi's *Anecdotes of Johnson* that should be valuable to teachers as well as parents:

I will relate one thing more that Dr Johnson said about babyhood before I quit the subject; it was this: That little people should be encouraged always to tell whatever they hear particularly striking to some brother, sister or servant immediately, before the impression is erased by the intervention of newer occurrences. He perfectly remembered the first time he ever heard of Heaven and Hell, he said, because when his mother had made out such a description of both places as she thought likely to seize the attention of her infant auditor, who was then in bed with her, she got up, and dressing him before her usual time, sent him directly to call a favourite workman in the house, to whom she knew he would communicate the conversation while it was yet impressed upon his mind. The event was what she wished, and it was to that method chiefly that he owed his uncommon felicity of remembering distant occurrences and long past conversations.

We need not press the anecdote too hard for a moral. Dr Johnson never had much difficulty in communicating any conversation, but it cannot certainly be said that he excelled in listening to others. Nevertheless, the practice described in the story is a discipline in listening as well as in communicating, and its application to school work is plain. No boy can repeat what he has not heard.

If a boy knows that, as a regular part of his work, he may be called upon to repeat to the class something that has recently been told him, he will also know that a regular part of his work is to listen carefully. The old 'code' composition exercise for Standard V, in which the children were required to write the substance of a short story read to them, may have been ineffective as a means of teaching the art of writing, but it was effective as a means of teaching the art of listening. Read a short simple poem to a class, even of older children, such a poem as "Home they brought her warrior dead," and ask individual children to come out and repeat it to the others, either in the original words or in words of their own, and you will be astonished to find, not merely how little they have remembered, but how little they have heard. Dictate to a seventh standard a short stanza—say, of *The Ancient Mariner* —and ask the boys to write it down after one hearing, and you will find that very few of them have really heard it. They have heard a few isolated words, but they have not heard the stanza as a whole thing The limiting of dictation to a mere spelling test is a misuse of a valuable instrument. There should be a regular use of sense-dictation to give practice in the art of consecutive listening. Dictate a stanza, not phrase by phrase, but as a whole, and expect, not mere accuracy of spelling, but accuracy of appre-hension. Dictate short prose passages—proverbs at first, then brief sentences, then longer sentences, and expect that the class should be able to reproduce, not merely some of the words, but all of the meaning. You will be training them to do something really difficult and really valuable, namely, to listen consecutively and constructively. There is far too much monologue in the teacher's school-work and far too much passive listening in the child's; frequent exercises in cooperative listening are good for both.

The many stories told of children's misapprehension of prayers and hymns conveyed to them orally indicate the need for some insistence upon sensible listening. "Old Father Whichard in heaven," "Lead us not into Thames Station," "For Emma and Emma," "Pity mice and plicity," and so forth, are amusing as blunders of childhood, but not as misunderstandings of maturity. There are people old enough to know better who apparently take no trouble to understand what is said to them. They have never learned to listen To know when not to listen is a valuable art in this world of bores; but the act should be under our own control.

IV

Systematic training in the art of writing

The person who cannot make a correct, concise, lucid and intelligible statement in writing is an uneducated person. Writing is not one of the great technical arts like the composition of music or the painting of pictures. Behind the *Chromatic Fantasia and Fugue* of Bach and the *Bacchus and Ariadne* of Titian lies an enormous technical equipment Behind the power to play the first or copy the second there also lies a great technical equipment. The art of writing demands no such quantity of gymnastic, but it demands a good deal One general impression is that everybody can write without taking special pains; but that is quite untrue, as we can discover daily. Many people are obliged to write letters, or to make written statements, and very few can do either The teacher's obligation is plain. It is still the fact that true ease in writing comes by art, not chance, even though the eighteenth century and the appreciation of form are both out of fashion The English view of writing oscillates between these two extremes, first, the belief we have already mentioned, that any one can write without special teaching, and next, a belief that no one can be taught to write by any quantity of teaching. Neither belief is sound. A few favoured persons are born writers, able to transmit their thoughts and their personality without effort; though even here we seldom know how much labour lies behind apparent ease. When a writer rashly confesses to taking pains in the practice of his art, as Stevenson did in an unhappy moment, many people regard the pains as reprehensible, the skill as suspicious and the confession as improper. Nevertheless, though no teaching will turn Tom Brown into Sir Thomas Browne, Tom Brown can be taught a workmanlike skill in the use of his own language, and, without teaching, it is highly improbable that he will ever acquire that skill. At the moment, Tom Brown is not taught to write and, as everybody knows, he cannot,—though I must add at once that increasing pains are being taken to teach him. He must, as a rule, be taught English through English. When Professor Saintsbury speaks of "translation and re-translation from and into

4—2

Latin" as "undoubtedly the surest (if not the only) way to master English writing," he is saying what is not helpful and what is simply not true; for it is painfully obvious that many thousands of English boys have spent several years in translating and re-translating from and into Latin without betraying the faintest symptoms of an immediate or subsequent mastery of English writing. This point is worth further consideration, for Mr Max Beerbohm, in the act of saying something entirely to our purpose, concludes by recommending the same panacea:

We have to consider that English is an immensely odd and irregular language, that it is accounted very difficult by even the best of foreign linguists, and that even among native writers there are few who can so wield it as to make their meaning clear without prolixity—and among these few none who has not been well-grounded in Latin.

I regret the final exaggeration because it may make readers doubt the entire truth of the rest. That English is a most difficult language to use lucidly, correctly and concisely is literally and absolutely true; but that no one can so use it who is not well-grounded in Latin is only partly true. Bunyan, who was not well-grounded in Latin, wrote a more classical English than Bentley, who was. Dickens, who was not well-grounded in Latin, is a more correct and shapely writer than Thackeray, who was. Jane Austen, who was not well-grounded in Latin, wrote a more exact and concise English than Scott, who was. Mark Twain, who was not well-grounded in Latin, wrote a much finer-edged English prose than Swinburne, who was. Mr Bernard Shaw, who was not well-grounded in Latin, writes as keenly and concisely as Mr Max Beerbohm, who was. On the other hand, many whose classical grounding is indubitable have written atrociously. Examples gross as earth exhort me, but I will quote one only, the deathless Henry Riley, B.A., of Clare Hall, Cambridge, who provided the old Bohn Library with translations that are a perpetual joy:

> Ipse miser vidi, cum me dormire putares,
> Sobrius apposito crimina vestra mero.
> Multa supercilio vidi vibrante loquentes.
> Nutibus in vestris pars bona vocis erat.

Thus Ovid.

Now hear the well-grounded Riley:

To my sorrow, in my sober moments, with the wine on the table, I myself was witness of your criminality, when you thought I was asleep. I saw you both uttering many an expression by moving your eyebrows; in your nods there was a considerable amount of language.

Once more:

> Di melius, quam me, si sit peccasse libido,
> Sordida contemtae sortis amica juvet!

Now hear the delicate Riley:

May the gods prove more favourable, than that if I should have any inclination for a faux pas, a low-born mistress of a despised class should attract me!

(I hope my examples will not be frowned at by the austere.)

Surely these passages are a sufficient answer to the claims of Mr Saintsbury and Mr Beerbohm.

The worst of such utterances is that there is just enough truth and just enough fallacy in them to make them dangerous. The fallacy is this: Mr Saintsbury and Mr Beerbohm had a natural instinct for letters, and they were, therefore, both consciously and unconsciously interested in such of their studies as provided a definite discipline in the art of writing. What they learned in Latin was naturally applied to English; but it need not have been Latin, it need not have been a foreign language at all. Any boy with an instinct for writing will take to the technique of his art, just as a boy with an instinct for music will take to the technique of his art. Stevenson, whose Latin was negligible, found his technical exercises in English; if he had been at Charterhouse with Mr Max Beerbohm the *College Magazine* essay would have sung the praises of Latin. Now what happened to be true of Mr Saintsbury and Mr Beerbohm is not true of boys in general. In effect these authors say, "We learned to write English from the discipline of Latin prose"—and so far they are partly right; but they continue, "Therefore no one can learn to write English without the discipline of Latin prose"—and there they are entirely wrong, both positively and negatively; positively, because, as we have seen, some excellent writers of English have been innocent of Latin, and negatively, because, not merely many, but the majority, of boys laboriously grounded in Latin, simply do not learn how to write decent English. That is not a matter of opinion, it is a matter of fact, and the theme of constant complaint. Our friend Riley had clearly no instinct for letters and his Latin did not make him write good English. No foreign language would have done so. What the Rileys need is not Latin or Greek, which their literary process is too feeble to assimilate, but sound, steady instruction in English itself; and that is what, as a rule, they never get. The general truth in Mr Saintsbury's assertion is just this, that an intelligent knowledge of Latin is of great advantage to the practised or practising writer of English. But surely that is a truth so obvious as to need no proclamation.

Mr H. G. Wells has recently been saying something on this matter, and his remark is specially worth notice, for the first demand he makes from the schools is precisely that which is urged in the present pages:

First we shall want our pupil to understand, speak, read and write the mother tongue well. To do this thoroughly in English involves a fairly sound knowledge of Latin grammar and at least some slight knowledge of the elements of Greek. Latin and Greek, which are disappearing as distinct and separate subjects from many school curricula, are returning as necessary parts of the English course.

We need not discuss this sentence in detail. Its claim for the classical languages is clearly not the same as Mr Saintsbury's, for Mr Saintsbury, I imagine, represents the kind of person who would die in the last ditch rather than consent to the dethronement of classics from supreme sovereignty to a mere servitorship in the realm of English. Classics like politics will always have its Jacobites. But what I want specially to say at this point is that, however helpful Latin grammar may be to the older writer (and no one denies its value), it is of no use whatever in the early or elementary stages of English. This perpetual harping on Latin whenever English is mentioned is a positive impediment to good teaching. We cannot say too clearly that the vast majority of English school children must be taught English without reference to any foreign language whatsoever. Think of the actual facts, and forget Mr Wells's abstractions. Here are our hundreds of thousands of young children, half-articulate, unable to speak or write, and surrounded by constant examples of debasement in language. We certainly want these pupils "to understand, speak, read and write the mother tongue well." But how, precisely, is Latin grammar or rudimentary Greek to help them? While children are engaged in unlearning their bad English it is fatal to multiply their confusion by the complications of any foreign tongue. Writers on education must endeavour to forget (or perhaps, rather, endeavour to remember) the circumstances of their own childhood when they make prescriptions for the elementary schools. What is true of upper or middle class boys is not true of all children. We must take elementary school children as they are. I have no hostility to the classical languages. Indeed, if I were the autocrat of education I would prescribe that the first foreign language learned by all English children should be Latin, partly because of the instructive value of the mechanism, and partly because, in the long run, Latin is the most generally useful language to the greater number of people. But that does not mean I should insist on Latin as a means towards the teaching of English. In any well-ordered educational system English would be the means towards the teaching of Latin, because by the time a boy began to get his grounding in Latin he should have been well-grounded in the elements of English. Without that grounding in English he will never, never learn Latin to any advantage. The Saintsburys and Beerbohms, born with an instinct for letters, are special cases, outside the argument. The fact we have to grasp very clearly is that all but a very few English children must learn to write English from English, and from English alone; and that to talk of the necessity or the advantages of Latin or Greek in this connection is merely to add confusion to difficulty. Let us repeat that, in general, the art of writing English must be learnt from English, and cannot be learnt except from English. The foreign auxiliaries will help, especially in later years, but they are only auxiliaries.

Many excellent masters can teach, and do teach, good English through Latin or Greek or French or German; but that is because they are excellent teachers, not because they teach foreign languages; and to talk of this foreign route as "the surest (if not the only) way" is, *pace* Mr Saintsbury, to talk nonsense—just the kind of nonsense that, on the one hand, has impeded the sane teaching of English, and, on the other, impelled the rough rude critic to denounce the classics as humbug. After all, people really know that the classics do not do what is claimed for them. This affectation of making every accomplishment in English depend upon some other language is an evil tradition. It is a form of snobbery, and has produced the infuriating Englishman who never really knows anything till it is told him in a foreign language.

In the elementary schools there is an evil tradition of another kind. A reference to page 19 will show that, under the old Code, children never heard of composition till they reached Standard V (about eleven), when they were expected to reproduce the substance of a short story read aloud to them twice. When they reached Standard VI a year later, they were supposed to have acquired the power (no one knows how) of producing essays on "Procrastination is the Thief of Time" or "Unpunctuality is the Mother of Invention." From these productions, the pronoun 'I' (the only one we have any right to use) was rigorously banished. We do better than that now; but a composition lesson, in a painfully large number of classes, still means that a subject is set and that boys write for some fifty minutes, while the teacher marks dictation or arithmetic. Subsequently, in the intervals of other work, or in his own leisure, he marks the compositions laboriously, correcting the more hopeful mistakes, or crossing out in despair the frequent passages that defy correction. This ceremony is gone through, once, twice, even thrice, every week. It can be described briefly as a hideous sacrifice of precious time and effort. The only compositions that can be corrected are those that least need correction. No one can correct a really bad composition. Methods of teaching composition need not be described here, for there are now several good books for teachers and several good volumes of class exercises. One or two broad principles may, however, be stated. The younger children should be encouraged to write and to write copiously—if they desire to write at all. Writing, as a regular exercise, should never be forced upon lower standard children. The notion that what they write is to be looked at censoriously and discussed disagreeably should never enter their little heads. The assumption must be that it is splendid and that teacher will love to read it. The teacher who can get little boys and girls to pour themselves out freely on paper may be well satisfied. Some very general corrections can be intimated, as for instance, that no really nice boy ever says "We was going up the Park," or "I never done it," or "Him and me went to the pictures"—though the teacher must sorrowfully prepare to accompany boys "up the Park" and to meet "him and me" as grammatical subjects, for many a long day; but any disheartening personal correction should be avoided. As the boy gets older, some attempt must be made to discipline his fluency into form. He has to learn that certain ways of expression are allowed and certain ways of expression are not allowed. Don't imagine that this will cause him surprise or distress. He will hear as calmly that there are good and bad ways of writing as that there are good and bad ways of walking and sitting and eating. He is already resigned to the dreadful mystery that things alleged to be good for him are often more disagreeable than those alleged to be bad, and that the easy way of eating is not the approved way; and he will accept the same sanctions and prohibitions about sentences without demanding the theological support of grammar. If there is one thing more pleasing and wholesome than anything else in the human boy, it is his entire disregard of first principles and his refusal to behave (in bulk) like the hypothetical Child of the educational treatises. That is what troubles young teachers: they have been led to expect The Child, and they encounter children. No one can or should lay down rules about the age or class at which the specific teaching of composition ought to begin. Every practised teacher will know by instinct when it is profitable to intimate instruction or to take specific lessons on the framing and linking of sentences. Roughly we can say that a little (but not too much!) occasional instruction can usually be given in Standard III. As the boy gets older he will need frequent exercises in the use of words, phrases, sentences and so forth. These are the scales of composition and must be practised very thoroughly and regularly.

Boys should not be expected to write more than one set piece a week; but what they begin must be finished, though not necessarily at one sitting and not necessarily in school. The practice of allowing boys to use a week's composition lessons in beginning half-a-dozen fragments of composition which they never finish is both specifically and generally harmful. Occasionally a boy may be allowed to scrap a failure—a very valuable experience for him; but he must recognise why it is a failure, and why it would be better to make a fresh start. Otherwise the rule must be that a composition begun must be worked out to its conclusion. It is a piece of valuable training for a boy to learn that his written productions, like the building he erects with his toy bricks, or the bridge he builds with his Meccano strips, or the things he makes with tools at his Manual Training Centre, must have substance, form and balance, must have too, a beginning, middle and end. The three stages of an essay or story or other composition should be always in the teacher's mind and sometimes in the pupil's: they are (1) Invention, or the assembling of possible matter; (2) Arrangement, or the selecting from the assembled matter; (3) Expression, or the embodiment of the selected matter. Of course no one will imagine that these are three dissociated processes; but they must take place and can be considered separately. A written composition is really a very severe test of capacity. To be confronted with a topic like A Loaf of Bread, or How to Lay a Table, or A Dog's Day; to be expected first, to assemble knowledge and notions; then to make an intelligent sequence of this knowledge; then to find adequate expression for it, is to receive a painful and paralysing intimation of one's mental emptiness. To work a set of sums is both easier and more mechanical! I should like to ask teachers how often since their own examination days they have been subjected to the literary inquisition they inflict so regularly on children. I shall return to this point later.

It is, of course, quite absurd to demand that all boys in a class must be writing on the same subject at the same time. It is just as absurd to demand that all boys in a class should be working the same kind or set of sums at the same time; but it usually is demanded. This ridiculous synchronism, this assumption that all boys are born equal (though not free) is one of the many evil results of the large classes that prevail in elementary schools. The public or secondary school master teaches a manageably small class. The Manual Training Instructor is not allowed to teach more than twenty. The unfortunate elementary school master is expected to teach sixty! No man can teach sixty children. He can lecture them or drill them, but he cannot teach them. In a lecture-lesson or drill-lesson all the pupils are supposed to be doing the same thing at the same time. In a practice-lesson they should be working at their natural speed; but where the class is large it is physically impossible for the teacher to keep in touch with all these varying speeds. Consequently in a large class every lesson tends to become a drill-lesson—the mechanical repetition of set exercises. For his composition exercise a boy has a reasonable right to choose the subject—so has the teacher; and justice should be done to both. Each must give and take. When A has finished his piece he should be allowed to begin another without any reference to the quantity written by B or C. The starting of a composition exercise, so to speak, by pistol-shot, with the understanding that all the runners must get to the tape in much the same time (without handicap) is a damnable inheritance from the tyranny of examinations. That a composition has to be written within a fixed limit of time is an accident of certain circumstances, not an essential of the exercise.

A composition must obviously be 'about something,' and have relation to the writer's knowledge, experience or suppositions. Composition is not like mechanics—it cannot be divided into 'pure' and 'applied.' There is no such thing as 'pure' composition. Even the most ethereal of lyrics or the most airy of literary fantasies (both utterly beyond the normal person's capacity) will prove to be 'applied' composition of an elaborate kind. On the whole it may be said that teachers do not make the composition exercises practical enough. Composition is too rarely applied to the matter of lessons, and, paradoxical though this may seem, too rarely applied to matter of fact. In the back of his mind the teacher will usually be found to have one standard for a set composition and another for an answer to a history or geography question. But explicit answers to clear questions are much the more valuable form of composition in school, if only because they have a meaning to the child that a set theme has not. They are certainly simpler for all concerned. The teaching of assembly, selection and expression will be found to flow more profitably from the statement-composition than from the inventive-composition, because in the former all the pupils will have been trying to write much the same sort of thing. If you ask a set of boys to write something about the Peasants' Revolt, theoretically they will reproduce the facts of the history lesson and their answers will be alike; but if you give them a theme, such as "When I was a Rebel with Wat Tyler," their opportunities for personal divagation will be infinite. Now it is much easier to base a lesson in matter, form and style upon the papers that are alike than upon the papers that are unlike. From the latter, indeed, many delightful and stimulating lessons may be drawn; but technical instruction in composition can more easily and obviously be applied to the more mechanical exercise.

So far I have written with the usual school subject called 'composition' in mind. But I have grown into a conviction that the usual school subject called 'composition' is a mistaken and useless activity, wasting more time and effort than any subject in the curriculum—except, perhaps, arithmetic. I will crystallise my views in a sentence and say, "Let us abolish composition from the curriculum." Certainly I hope I shall live to see composition as it is usually understood, consigned to everlasting perdition with (let us say) mental gymnastics, literature (as she is taught), grammatical definitions, parsing, model-drawing, and problems, as the seven devils of the school-room. As a provisional measure let us cross out 'composition'—that word of ill omen—wherever it is mentioned in time-table or syllabus and put some description like 'Constructive English' in its place. If any reader thinks I am going to extremes let him ask himself these questions: Do children like composition? Do children ever ask to be allowed to write as they ask to be allowed to draw? Do their faces brighten when you say, "We are all going to write a composition this afternoon"? I will assume the negative answer and say that the reason why they do not like composition is that composition is something entirely beyond their power. The 'composition'—the essay, the impression, the 'chose vue'—is really a difficult and recondite form of writing. Consider how few first-rate essayists there are, consider how few are the essays that give permanent delight, and you will recognise how foolish is the attempt to turn the creative faculty of children into this narrow and precarious direction. I said on an earlier page that we cannot turn Tom Brown into Sir Thomas Browne; but that is just what we have all been

trying to do! We seem to proceed on the assumption that
the boys' filled exercise books must resemble *The Essays of
Elia*, and our teaching is bad because we dimly recognise that
there is a serious mistake somewhere—that we are laboriously
exercising the pupils in a difficult, artificial gesture, that can never
form part of their normal, natural life. I asked above how often
teachers had subjected themselves to the composition tests they
inflict on children. I will push the question to the point of rude-
ness and ask how many teachers themselves could write the
compositions they demand from children? How many teachers of
composition can be called connoisseurs of composition? If they
are not, it is dangerous for them to presume to teach something
beyond the range of their own accomplishment or appreciation;
if they are, the more hopeless they will find the task of teaching
a delicate and difficult art. The root of the trouble is surely this,
that we have not clearly distinguished in our own minds between
the writing that is a statement or record, and the writing that is
a creation or invention. What we ought to be trying to teach is
the first; what we are usually trying to teach is the second; and
the vital fact is that, whereas the first can be taught, the second
cannot. I hasten to add that I should never dream of ruling out
attempts at literary invention from the school curriculum. On
the contrary, I would encourage them; but I would encourage
them in more profitable directions. What I demand at the moment
is that we should ruthlessly clear our own minds about this time-
honoured school subject called 'composition.' Analogies are not
argument, but they are sometimes instructive. Music is taught
in most schools. Well, I ask teachers if they would ever dream
of giving out sheets of music paper to a class, once a week or
oftener, with a demand for a song, or a study, or a prelude, or
an impression, or a musical composition of any sort; and I ask
further whether the practice of giving out sheets of writing paper
to a class, once a week or oftener, with a demand for a story or a
sketch or an impression or a dialogue or a literary composition
of any sort is not very nearly as absurd? In music we can reasonably set exercises in tune and time and rhythm; but we cannot
reasonably make demands for composition. In writing we can
reasonably set exercises in vocabulary, construction and statement: but we cannot reasonably make demands for composition.

Teachers must not imagine that, because everybody uses words in daily speech, creative art in the medium of words is easy or possible to everybody. The art of writing is difficult. Some great writers to the end of their days have never found facility, but have had to wring out words by the sweat of their brow. To set a piece of blank paper before a child and expect him, first, to spin out of a scarcely existent self something that is not there, and next, to present this something in a form that is technically correct, is to expect a miracle—or, to speak without rhetoric—it is to make a demand that is disheartening, disabling, unjust and almost cruel. In a recent examination for the Junior County Scholarship set by the London County Council, this was one of the two questions that constituted the English paper.

"Suppose that Shakespeare and Nelson met to-day in London. Write a conversation between them." (The candidates at this examination must be less than eleven years old and working in Standard IV and above.)

Now what can a child of ten-and-a-half in Standard IV (or above) know about Shakespeare—or about Nelson? What kind of conversation can such a child possibly invent appropriate to a meeting between an early sixteenth century character of which he barely knows the name, and an eighteenth century character of which he knows little more? How many teachers could write a satisfactory composition on such a theme? Does not the question amount to a demand for creative literature of a singularly delicate and recondite kind? You may think this an extreme view to take of a question set at a half-yearly scholarship examination; but that examination is, in effect, an examination of all the schools by sample, and it *is therefore an examination that is prepared for*. And thus our ten-year-olds in Standard IV (or above) have to spend their poor little powers in super-Landorian exercises of this sort. I say roundly that it is worse than preposterous to inflict such exercises upon children, and I say further that from no quantity of such exercises will our children ever learn the rudiments of composition. When I demand the abolition of composition, that is the sort of thing I want abolished. I shall doubtless be told that nowadays we do not expect our children to write essays and that the whole subject of composition has undergone a revolution. I can only reply in the words that

Alphonse Karr applied to another kind of revolution, "*Plus ça change, plus c'est la même chose.*" We really have not changed much when we ask the child to become Landor or Lucian instead of Lamb or Bacon. Composition, by whatever name we call it, comes to much the same thing in the end. We cannot, I say, reasonably give a boy a piece of paper and order him to go and create something; but we can reasonably order him to go and record something. Creation is his affair, not ours; but statement, record, reproduction—the mechanical art or craft of writing, that is our affair. Few of our pupils will be called upon to become literary artists; but most of them will have to make statements in writing. What, therefore, we must do is to teach children how to make brief, clear, correct and concise statements.

Teachers themselves should have undergone a precisely similar training. I have asked for the abolition of composition in schools, I ask also for the abolition of composition in Training Colleges —understanding by composition the writing of set Essays. If teachers are compelled to commit the crime of formal composition as a regular part of their own training they will naturally urge it in turn upon their pupils. At present, the Essay is a compulsory paper at the Teachers' Certificate Examination. It should be abolished, and the marks for composition should be awarded by the examiners after a close scrutiny of the candidate's written work in such subjects as Literature, History, Science, Geography and Pedagogy. The student-teacher will then be made to understand that what is required from him is not the strained and artificial gestures usually found in examination essays, but the ability to make clear, correct, concise and well-ordered statements in plain workmanlike prose. Further, he will be made to understand that, as every teacher in English is necessarily a teacher of English, every teacher, without any exception, must possess that ability, or be, so far, disqualified as a teacher. Those who intend to specialise in music or drawing or needlework or cookery or carpentry can claim no exemption. If people are going to teach at all in any subject they must be able to express themselves clearly, cogently and correctly, without awkwardness, affectation, weakness, inflation, slovenliness or any other offence against good taste. No one will expect teachers as a body to be artists in words; but no one must expect them not to be capable artisans. Honest,

careful craftsmanship in language—that is what we must demand
from the practical composition of all teachers. I suggest, however, that the Certificate Examination should include an *optional*
composition paper to be taken by those who are genuinely attracted
by the art of writing, or who have any creative leanings, and that
the marks gained in this paper, together with the general composition marks, should count towards Honours or Special Distinction in Composition. It would have to be made clear that
no marks at all would be awarded to Composition papers not
reaching a very fair standard of excellence; for we do not want
to abolish compulsory composition merely to find it coming back
as optional composition attempted by everybody. We want
students to understand that there are two kinds of writing, the
statement and the creation: that ability to state is expected from
them, but not ability to create, though we are ready to give them
credit for genuine attempts at creation. At present they do dimly
understand that there are two kinds of writing, but their understanding takes the form of a vague idea that, though you must
write with elaborate and uneasy ceremony in the paper called
Composition, you may write anyhow in the paper called History,
as long as you are careful not to say that King Alfred shot a burnt
cake from the head of his little son Prince Arthur outside the
Banqueting House at Whitehall. What we must make students
clearly understand is that we do not want the elaborate ceremony,
and that we will not tolerate anything written anyhow—that, in
short, everything they write is 'composition.' When they understand this they will give up the attempt to teach their pupils
composition and begin to teach them how to write. The lessons
in history, geography, science and kindred subjects will afford
abundant material, both for expanded and summarised statements.
Merely to ask boys and girls to set down in writing how they
would clean a pair of boots, or how they would lay a table is to
give them really efficient practice in the craft of writing. We do
not want them to be literary, we want them to be precise and
practical. To ask them to pretend they are the Moon writing a
letter to the Sun is to expect them to be literary before they are
articulate. Here someone will say he has actually seen charming
letters to the Sun written by little girls of ten in the Junior
Scholarship Examination. I do not doubt it, but I ask, first, what

proportion the charming letters bear to the mass of school composition that is not charming, and next, whether exercises of this sort should be juvenile human nature's daily food in the way of composition. By all means let the pretty fancies of children have their way. Encourage them, and encourage the expression of them; but do not imagine you are going to secure in the general mass of children a workmanlike command of their language by daily demands for these profuse strains of unpremeditated art. Our business, so to speak, is with the sparrows, not the skylarks.

It is towards the story and the play, and not towards the essay or 'composition' that the creative activity of children can be most profitably directed. For purposes of school work a play has three aspects, it is something to be written, something to be read, something to be acted. The last two aspects do not concern us at the moment; but let us consider the first. A play, as something written, may be original or adapted. The plays of J. M. Barrie are examples of original drama, and the plays of Shakespeare examples of adapted drama. These obvious instances are given as a reminder that the attempts of school children at either line of composition have a parallel and justification in the realm of public art. The writing of plays in school is a very valuable and practical form of composition—much more practical than the writing of essays. In a sense, children are primitive beings, and the essay is not a primitive form. The essay has no place in classical literature, and none in modern literature before Montaigne. The bison painted in the cave at Altamira indicates the existence of high artistic capacity in ages almost immeasurably distant; but there is no evidence that the men of Cro-Magnon took in each other's essays. Epics existed before essays; the world had a large body of narrative and dramatic literature before it arrived at the essay: and yet it is precisely this difficult and fragile—even sophisticated—form of composition that our juvenile pupils are expected to produce! The essay for them has no purpose. Children know what a story or play is long before they know what an essay is. They can understand writing a story for the class magazine, or a play for a class performance; they cannot understand writing an essay for the waste-paper basket.

The collective composition of a play can be attempted by quite young pupils. As soon as boys are old enough to enjoy a ballad or a story in verse they should try to translate it into action. Thus, the defence of the bridge by Horatius and his comrades is a fine exciting story that can be done in action. But action without words is only half the fun. The class will, therefore, have to fit words to the action—that is, translate the narrative of the poem into drama. They must decide the point at which they will begin, the speaker who is to open the scene and the words he is to utter. The sentences approved by the class will be written down by the teacher (who is merely the humble scribe) and when something like a scene has been achieved it can be tried over and its short-comings detected and corrected. The one scene can be expanded to a couple or more as the exigences of the story demand, and so a play is made. A familiar story or fairy tale or a famous historical incident can be treated in the same way. An older class will not be content with a simple scene or two of which the plan is more or less ready made, but will like to invent a drama of its own. History is again a fruitful source. Consider the training involved in the composition of a drama on the subject, say, of Sir Walter Ralegh. There is the work of planning the whole drama, then of planning each scene, then of choosing the characters, then of fitting the characters with becoming words, then of making the scenes accord with the conditions of time and space—of time and space in the artistic, historical sense, and of time and space in the practical, theatrical sense. Then there is the trying over, the correction, the expansion of this, the excision of that, and so on. All this is training in the writing of English such as the weekly attempts at a 'composition' will never give. It is in the fullest sense practical English composition.

The collective composition of a story is an equally excellent and exciting adventure for a class. The pupils must first decide what kind of a story it is to be—a school story, an adventure story, a historical story—and what its general course is going to be; they must choose their principals (the minor characters have a habit of coming in by necessity rather than by choice) and then embark on the opening chapter. They (usually a few individuals) will suggest the sentences, and those approved by the class will be written down by the teacher. The class will soon see if the story works or if it halts and breaks down.

One immediate effect of this joint composition is usually an outbreak of individual composition—which is just what we want. And there should be a place for the best efforts, namely the Class Magazine, of manuscript paper (with drawings intermingled), produced by the boys themselves. No class of twelve-year-olds or above can be considered alive that does not venture on a magazine. As a means of creating 'class consciousness' (of the right kind) a class magazine is invaluable.

When a boy wants to try his pinions alone he should receive encouragement and sometimes tangible support. Let his reading help his writing. For, consider, what mental stock has a boy of twelve or thirteen? If he is an elementary schoolboy, that is, a boy from a poor and probably bookless home, what general mental stock can he possibly have? What has he seen or read or heard or done? He has had no adventures, has rarely been away on holidays, and has spent in the streets near home what leisure has been left from school and errand-getting and odd jobs. Nowadays he has 'the pictures,' and instead of cursing them solemnly as most pastors and masters seem to do, we should bless them as a tremendous means of opening and stocking his mind. To ask the poor elementary boy to express himself in writing is like asking a penniless man to be liberal with his money. The spirit may be willing, but the means are to seek. You are really asking the boy to give you what he has come to get from you, namely, ideas, notions and views of life. If the boy wants to write, let him begin with the help of his betters. Give him something to imitate. Most schools possess some Dickens volumes. The paragraphs of Dickens are models of structure and punctuation, and can be set as examples for imitation. As a boy I always liked that old composition book by Dr Theophilus Hall—I enjoyed the specimens of style he gave, and I enjoyed the hopeless task of trying to write something to the patterns set. I mention that kind of exercise without recommending it for general use. Elementary school children are, as a rule, too young for it. At the worst it is a more valuable exercise in writing than paraphrase, for, rightly used, a paraphrase is a test of reading rather than of writing—the implied question being, not, Can you write this passage in a different way? but, Do you understand what this passage is about?

It should be unnecessary to say that no one can teach composition properly without a trained and cultivated sense of writing; unfortunately it is not unnecessary[1]. Teachers must keep their prose conscience active if they presume to teach others how to write.

As I must suppose that 'compositions' will continue to be written, I will add a few remarks on correction. Most of them will apply to the simple exercises in construction and statement that (in my opinion) should replace the cherished essay. The correction of any written work is necessarily a laborious task, even though the red ink interlining be abandoned. Every piece should be read by the teacher and have a mark assigned to it. The standard should be high. The full mark should be reserved for exceptional merit—the V.C. of composition. Elaborate correction of individual books is quite useless. What is wanted is collective correction or criticism. Let the teacher take one or two typical examples (of excellence or mediocrity or failure) and initiate a class discussion. The trouble will always be to make the critics direct their attention to the major qualities of matter, form and style; they will generally devote their ingenuity to the detection of minor, and sometimes imaginary faults; but as the major qualities of writing are precisely what the teacher is anxious to exhibit, the trouble, however great, is well spent. A correction book containing the typical faults of a week's batch of pieces will give invaluable matter for lessons or discussions. It will show the prevalence of certain errors and indicate the direction in which special efforts at correction must be made. No teacher should make the mistake of expecting or demanding uniformity of style or treatment in the writing of a class. *Tous les styles sont bons, hormis l'ennuyeux* says a Frenchman. Honest sincere thought should be encouraged, and windy platitudes discouraged; but teachers should be tender with the child who likes to be a little 'flowery'—it is natural for the young to be artificial! The one

[1] The necessity can be illustrated by an example. An exercise book used by an adult student for one teacher's work contained, on a certain occasion, a composition written for another teacher. At the end of that strayed composition exercise the teacher of composition wrote this: "Please keep my work in a different book to the other teachers; it adds confusion." It would be difficult to achieve more mistakes in a dozen words.

really hateful style is that based on the jargon and clichés of newspapers and offices; and this unfortunately, is a style that is only too easy to acquire. One recent prize essay in a competition was composed almost entirely of journalistic clichés and inflations. Nor do we want a schoolmaster's English. There is a kind of discount usually known as 'banker's discount,' and another kind called by commercial people 'schoolmaster's discount.' Schoolmasters themselves call it 'true discount.' They would. There is also a kind of elaborated uneasiness in writing that we may call 'schoolmaster's English.' Let us avoid it. It is a sham. If we are to ask for sincerity in our pupils, we must first be sure that we really appreciate it ourselves. A class of adult students (they were teachers) once elected to write a composition on Wild Flowers. Most of them began after this fashion: "Beautiful as are the blooms that adorn our gardens in summer, they are surpassed in loveliness by the little wild flowers that grow unregarded in the woodlands and meadows." When I denounced this as humbug and asked if they really thought that any wild flowers were lovelier than our garden roses and carnations, they cheerfully and readily replied that they did not, but that they said so because they thought it was the kind of sentence they were expected to write in an essay! In their answer it seems to me that the final damnation of the usual 'composition' has been uttered. What can teachers expect from pupils if they are not sincere themselves? What sense of artistic proportion can they cultivate in others if they lack it themselves? A young friend of mine (aged just seventeen) at a well-known secondary school was required to write an essay on "The treatment of death in *Ecclesiastes*, *The Pilgrim's Progress* and *Urn Burial*"! At seventeen, "What should she know of death!"—what indeed, of *Ecclesiastes* and Sir Thomas Browne? Two young ladies, also in their teens, asked me one day where they could read something about the character of Falstaff. I suggested *Henry IV*, and they were visibly disappointed. They then explained that they had been set to write an essay on the character of Falstaff, and what they really wanted (though they did not put it quite so crudely) was a source (not easily spotted) from which they could lift a few choice sentences. Surely neither Sir Thomas nor Sir John—neither that spirit nor that flesh—is matter for girls of that age! If we stuff our pupils with humbug and hypocrisy how can we expect them ever to see the truth of great art?

The chief vices of education (says Ruskin) have arisen from the one great fallacy of supposing that noble language is a communicable trick of grammar and accent, instead of simply the careful expression of right thought. All the virtues of language are, in their roots, moral; it becomes accurate if the speaker desires to be true; clear, if he speaks with sympathy and a desire to be intelligible; powerful, if he has earnestness; pleasant, if he has sense of rhythm and order. There are no other virtues of language producible by art than these; but let me mark more deeply for an instant the significance of one of them. Language, I said, is only clear when it is sympathetic. You can, in truth, understand a man's word only by understanding his temper. Your own word is also as of an unknown tongue to him unless he understands yours. And it is this which makes the art of language, if any one is to be chosen separately from the rest, that which is fitting for the instrument of a gentleman's education. To teach the meaning of a word thoroughly, is to teach the nature of the spirit that coined it; the secret of language is the secret of sympathy, and its full charm is possible only to the gentle. And thus the principles of beautiful speech have all been fixed by sincere and kindly speech.

How far do our 'compositions' and 'essays' go towards revealing the secret of language and the sincerity of great literary art? Are they not, in fact, what Stevenson called his youthful diaries, "a school of posturing and melancholy self-deception"? Sincerity is a moral quality, and we must demand that always. The technical quality to be insisted on above all others is lucidity. The ability to state clearly, shortly and exactly what we mean is very rare; and to teach that ability is a prime duty of those responsible for composition lessons. To go further is dangerous. We can teach children to write clearly; we cannot teach them to write beautifully, though we can make them aware of beauty.

Nothing has been said, so far, about grammar. My own conviction is that grammar as a science, as a subject-in-itself, with a name of its own and a separate place in the time-table, ought never to form part of the elementary curriculum. It is a study for the senior schools, certainly not for the elementary schools. It is really an adult subject; the administration of it to young persons is dangerous. For the elementary schools, grammar is simply an aspect of the writing and speaking of English, and should never be anything else. Grammar as a separate subject inevitably means definitions and difficulties. My ears still recall the voices of Standard IV boys filling the air with their sweet jargoning as they chanted definitions of relative pronouns, mood and prepositions—yes, actually those! And I remember finding a Standard VII engaged in analysing a dozen lines from *Henry V*, containing an anacoluthon. That is what always happens, when grammar is an isolated subject on the time-table. Grammar is, in fact, a science, and a difficult science, because inexact. But though grammar should not be taught in school as a thing-in-itself, it can be considered as a thing-in-itself, its existence acknowledged, and its name and rules invoked. Banishing grammar from the time-table does not mean banishing grammar from the school. In the course of composition lessons such matters as number, gender, concords, cases of pronouns, subject and predicate, transitive and intransitive verbs, object and the commonly used tenses must necessarily come up for notice and they can be shortly and practically dealt with in relation to the art of writing. Casually, incidentally rather than specifically, the junior classes will get to know the names of certain parts of speech, and the senior classes all of them. The senior classes should know the structure of a simple sentence and how a sentence may be enlarged by phrase and clause. The golden rule must be never to seek for difficulties. Deal only with the grammar of such English as the boys actually use in writing—there will be more than enough difficulty in that! Remember that some of the children will soon have to begin a foreign language and will need to be acquainted with the usual grammatical terms.

I have been rather short with grammar, but, even thus, I am afraid I may have said too much in its favour, because I know the tendency of teachers to go too far in either direction—I have met a teacher who would not so much as mention the words 'noun' and 'verb' to a class on the ground that 'grammar' was not taken in the school. Let me put my view with aggressive brevity and say that *it is impossible to have too little grammar at the elementary stage of education*. Certain grammatical terms—such as sentence, subject, object—are time-saving appliances and naturally must be used in teaching. To use a term, however, does not in the least involve an obligation to explain it. The besetting

sin of the conscientious, methodical elementary teacher is the
desire to explain everything, and to precede the use of a term with
complete and accurate definition. Thus, in his view, if boys are
going to deal with verbs, the first great step is to arrive at a sound
definition of a verb by the principles of orthodox pedagogy, and
the next step is to see that the definition duly achieved is committed to memory. Thus equipped the boy is fit to deal with
verbs. Now in actual life we do not learn anything by definitions.
As I said earlier, we all know a dog when we see one, though none
of us could define a dog. I am sure I know any verb when I see
one, but I am quite sure I don't know how to define one. In
teaching most things to children we must be pragmatists. Do we
need to talk about prepositions? Well, these are prepositions, *by*
the fire, *under* the table, *at* the war, *in* the soup. Children will
instantly understand what a preposition is by seeing it (so to
speak) alive, and doing its work; by definition they would never
understand a preposition to their dying day. We really must get
out of our heads the idea that a child learns things by starting
from first principles. I remember when I was an assistant master
in my first school and had to teach drawing, it seemed to me a
gruesome thing to afflict lively and eager boys with those unreal
and arid 'models'—cubes, cylinders, hexagonal prisms, and so on.
I therefore got some real things for them to draw, including a
large jam-jar. But my old head-master (peace to his soul, he was
a good friend to me) would have none of me and my jam-jars;
and I have never forgotten his remark—"when they can draw
a cylinder perfectly, then you can give them a jam-jar to draw";
—it has remained in my mind as the typical fallacy of 'first
principles' in teaching. This fallacy lies at the root of much
argument about the classics—for instance, that you cannot understand the English poets till you have read the classical poets who
were their source and inspiration. The same fallacy lies at the
root of our fond belief that grammar is useful. It is useful; but
not to children. Little Tom Brown of Prince's Gate talks
charmingly, but not because he is well-grounded in grammar;
little Tom Brown of Prince's Gate Mews talks badly, but not
because he has omitted to study the principles of grammar. Almost
the only common fact in their English is their total ignorance of
English grammar. They talk well or ill purely as a result of

unquestioning imitation. The amount of practical help a boy will get in speech or writing from grammar is infinitesimal. But the teacher himself should be clear about grammar; for the clearer he is, the less tempted he will be to inflict difficulties on children. A grammarian was very fierce with me one day because I denied that a boy was better off for knowing that 'go' in 'I shall go,' is in the infinitive mood. I still deny it; and I assert on the contrary that to trouble the juvenile mind with the infinitive mood is to make it worse off, not better off. But whatever our views on grammar in general I think most of us will agree that in the early stages grammar should be a means, and never an end in itself, that it should be reduced to the helpful and necessary minimum, and that it should be as simple as such a difficult subject can be. What happens when grammar is treated as a separate subject can be best illustrated by an example. Recently the girls in the Lower Fifth of a County Secondary School had to analyse this passage:

We can only have the highest happiness, such as goes along with being a great man, by having wide thoughts, and much feeling for the rest of the world as well as ourselves; and this sort of happiness often brings so much pain with it, that we can only tell it from pain by its being what we would choose before everything else, because our souls see it is good. (*Romola.*)

Really, it is nothing short of an outrage that girls of fourteen and fifteen should have such exercises inflicted on them; it is an outrage that such a passage as the above should be brought to their notice at all except for reprobation as a piece of bad writing; and it is certainly an outrage that crimes of this kind should be committed in the name of English. The exercise is so far from exceptional that the unfortunate girls have a book full of similar passages and have to dissect one each week. Surely no one will pretend that such exercises have any purpose, intellectual or emotional, useful or ornamental. The one effect they certainly achieve is to make the victims hate English with peculiar intensity. We must not have the shadow of this sort of thing in the elementary schools. I think it is fairly safe to say that the worst science teaching, or French teaching, or geography teaching, or indeed any other kind of teaching, is never quite so bad as the worst English teaching and never quite so common.

Personally I should like our bright top-class children in the elementary schools to attempt a little logic. It would be at least as useful as grammar, and there is just the touch of ritual (or rigmarole, if you like) in it that boys like I am sure that much of the value of Euclid lay, not in the geometry or the alleged gymnastic, but in the delightful rigmarole. Boys love rigmarole, and elaborate nonsense, and quote it at each other with glee. Now the rigmarole of grammar is the wrong rigmarole. There is no fun in it, as *The Comic English Grammar* painfully proves; but there is fun in the rigmarole of logic. Look at Lewis Carroll!

The question of spelling arises in connection with the writing of English. As we have already pointed out, some spelling mistakes can be traced to faulty enunciation. Other so-called errors of spelling—'there' and 'their,' 'as' and 'has,' are really matters of syntax; the boys know how to spell the words; what they don't know is when to use them. Spelling lessons in the main should be based on the children's own vocabulary, i.e. they should learn to spell the words that they want to use. Lists of the words frequently miss-spelled in the composition exercises should be collected by the teacher and specifically dealt with. No method of teaching spelling carries us very far; the anomalies of the language must be frankly faced as difficulties and dispatched as they arise. A boy should always have access to a dictionary as soon as he knows how to handle it, and should be encouraged to use a wide vocabulary in his composition. Regular exercises for the extension of vocabulary should be given. They are amusing as well as useful. Roget's *Thesaurus* is a gold mine of examples

So far we have spoken only of prose. The writing of verse should not be overlooked. It is remarkable how men who believe that the writing of Greek and Latin verse is a useful part of a classical education seem startled when it is suggested that the writing of English verse is a useful part of an English education. The two things are not quite the same, of course; but they are not entirely different. Verse-writing certainly has a place in the English course, but it should not be a place within the meaning of the time-table. It should be suggested to the boys, not imposed on them. When they find that the good verse maker is praised

and admired they will be stimulated to try. Quite little boys will
often produce some charming examples of *vers libres*. It should
be added that school verse-writing—and school play-writing and
school story-writing—should never become a 'stunt' or advertised
activity. Schools ought not to wash even their clean linen in
public. The place for the best class verse is the class magazine.

V

Systematic training in the use of books

The person who cannot make practical use of books as sources
of information is an uneducated person. Samuel Butler is alleged
to have said that no man needs more than two books, *Whitaker*
and *Bradshaw*. Certainly the person who can use *Whitaker* has
a vast store of information at his disposal, and our upper standard
boys should be made aware of its existence. I encourage such
boys to get for themselves a copy of *Pears' Cyclopaedia* (in happier
days obtainable for a shilling); there are few lessons in which they
cannot get practice in using it.
In general, the growing boy must learn how to get the marrow
from a simple treatise. He must learn how to read a history book
for its historical information, a geography book for its geographical
information, and so on. He must be able to use an ordinary
dictionary and such books as Smith's *Smaller Classical Dictionary*,
a copy of which should be in every classroom of Standard V and
above. He must learn how to use a simple encyclopaedia, a bio-
graphical dictionary, a gazetteer, an atlas, the index of a book, and
a library catalogue. And it will not hurt him if he learns that
there are such works in existence as the *Dictionary of National
Biography*, the *Oxford Dictionary*, the *Encyclopaedia Britannica*,
and the *Annual Register*. I should like to see a simple form of
Roget's *Thesaurus* for school use to help the young pupil in his
writing. It will be seen that we have left the limited province of
the English master and gone afield in the great Empire of
English—into the great world of books. It is a very necessary
excursion. The men, and especially the women, who are the

products of the elementary schools are generally a race without books. Their reading matter is current fiction in its cheaper forms, and even this is usually borrowed. Books other than cheap novels are to them a kind of furniture, like pictures—the sort of thing it is respectable to have in the parlour, but no more to be read than the Family Bible. Certainly our working classes (the product of the elementary schools) appear unable to use books either for pleasure or profit, and have no delight in possessing them.

Above all things (says the Autocrat), as a child, a man should have tumbled about in a library. All men are afraid of books, who have not handled them from infancy. Do you suppose our dear Professor over there ever read *Poli Synopsis*, or consulted *Castelli Lexicon*, while he was growing up to their stature? Not he; but virtue passed through the hem of their parchment and leather garments whenever he touched them, as the precious drugs sweated through the bat's handle in the Arabian story. I tell you he is at home whenever he smells the fragrance of Russia leather.

To teach the use of books should therefore be one of the school aims. Boys should be set little pieces of research involving some cross reference and the consultation of several volumes. In most London boroughs the children are fortunate in having a public library at their disposal; in remote rural districts it is an imperative duty of the local education authority to establish a small reference library in each school. During the short time we have our boys and girls we can do few things more useful than to show them how to use books as tools, as sources of information and as the means of further study. Perhaps if we can teach them how to get fact from print we may be teaching them not to draw opinions from print. The danger of the age is the perpetual vociferation of inflammatory opinion by all sorts of periodicals (and their placards) at all sorts of prices. If we can make our pupils understand that behind print, even of the largest circulation, there is merely a man of no special importance, we may be doing them and the world a lasting service. We want to create people who can use print, not people who are intimidated by print.

If teachers can be persuaded to forego some of their monologues and to substitute lessons in which the pupils have to get their own information from class text-books or volumes borrowed from the library they will save immediate time and be teaching the boy how to lay up almost the only kind of treasure on earth that no one can tax out of existence. I don't want to have the appearance of continual exhortation, but I must remind teachers again that one of their faults is the well-intentioned crime of doing too much. The ideal teacher is the one who does nothing while the class does everything; but the actual earnest teacher is usually someone who works himself (or herself) to exhaustion while the pupils sit silent, often bored, absorbing (say) about one-tenth of the information poured out upon them, and, incidentally failing to acquire one of the greatest gifts that school should give them, the ability to help themselves. Once more I beg teachers to remember that their work is elementary. They have merely to begin something. Teach the children how to use books; teach them, in fact, how to read; direct their reading, test their reading, and show them how to extend their reading, and you will do more for them than if you talked high wisdom to them sans intermission for five hours a day. The best thing a child can get from a teacher is not a mass of authoritative information that settles everything for ever, but the thrilling assurance that life is a great adventure worth pursuing in a spirit of discovery.

VI

The induction to literature

With the approach to literature we reach a different plane of work. So far we have considered English almost solely on its practical side—as the acquisition of necessary skill in speaking, writing and reading. We have dealt with the aspects of English that everybody must teach without exception; we now touch upon the aspect of English that nobody must teach without special ability to receive and transmit. We reach the English that is not a routine, but a religion.

The reading of literature is a kind of creative reception. It is almost sacramental. In the ordinary sense of the words, literature cannot be taught. If in any school something called literature is systematically taught, the efforts will usually be found to be directed towards literary history, or 'meanings,' or the explanation of difficulties, or summaries of plays and stories, or descriptions of characters, or chatter about Harriet—all of which may be useful, perhaps desirable and even interesting, but all of which are evasions of the real work before the teacher responsible for literature. It is, unfortunately, a familiar fact that a poem without 'difficulties' is for many teachers itself an insuperable difficulty; they cannot find or give any 'notes' on it.

The true teacher does not teach, but transmit. What the teacher has to do with lines like—

> Breaking the silence of the seas
> Among the farthest Hebrides,

is not to describe the Hebrides or give an account of Wordsworth's visits to Scotland, but to transmit the magic of the utterance, or, at least, to remove anything that may impede the transmission. There is a well-founded tradition in a certain district of London that an old elementary head-master, examining a class in *The Merchant of Venice*, and encountering the line "Came you from Padua, from Bellario?" called for a map and insisted that boys should point out the position of Padua, and then the position of Bellario. The story is supposed to be funny, but it is really tragic. Its real point is not the comic ignorance of the head-master, but the evidence it affords of the outrages committed in schools in the name of literature. No wonder children left school without regret, and shuddered ever after at the mention of poetry. It is still possible to use the present tense in that sentence

If literature in schools is not a delight, if it is not, in all senses, a 're-creation,' an experience in creative reception, it is a failure. The teacher is, as we have pointed out, not so much a teacher, as a transmitter. What has to be presented to the children is not a piece of information, or a 'lesson,' but a living creation, dateless, timeless, eternal, the moral of which is that it has no moral, and the lesson of which is the Rhodora's lesson, that beauty is its own excuse for being. For the work of presentation a pleasing voice and a responsive soul are more necessary than much knowledge and a research degree. Not for a moment would I seem to depreciate literary knowledge and research or the academic recognition of either. I am thinking of the special need and the occasion—the class of young barbarians whose souls are to be touched with the magic of poetry and whose souls will certainly not be touched unless there is first a soul to touch them. Academic distinction is warrant for a mind, but not for a soul; it does not pretend to vouch for an emotional equipment. Thus the task of choosing a specialist teacher for literature is peculiarly difficult. There is nothing to go on. University qualifications are a safe enough guide when you are looking for acquirements—when you want a science master or a history master; but not when you are choosing someone to be a medium for the transmission of the spirit. There is, of course, at least as much chance that a graduate in honours will be the right kind of person as that an untrained teacher will be the right kind of person; but there is no certainty. That is why it is absurd of education authorities to demand 'a good honours degree' from applicants: they are taking the guarantee of one quality to be the guarantee of another. The successful teacher of literature must have a little in common with the actor. In the slang of the stage he must be able to 'get it across' to the audience. If an actor is going to play Shylock, he must not come on the stage and give a lecture embodying his views on the text of the play, its construction, stage history, characters and so forth. He has to take his text and 'get it across' to the audience. So the teacher who is going to read *The Merchant of Venice*, or *The Ancient Mariner*, or any piece of pure literature, must take the text and 'get it across' to the class. That is not to say that the teacher is forbidden to talk about the play or the poem; he may do that with great advantage, if he can do it well; but it is altogether subsidiary to his real business, which is to pass on a living spark undimmed. Neither is it to say that the teacher must be an actor—heaven defend any class from the histrionic teacher! But the teacher and the actor are alike in having to depend upon something personal, something emotional, something that is not acquired knowledge or information. I should like to add from my own observation, and for the encouragement of beginners, that the teacher, like the actor, is made as well as born. He will develop slowly, but certainly—and develop not merely in technique, but in the personality of which technique is the instrument. The teacher may need almost the *douze années* that Talma thought necessary for the making of an actor; he will certainly be better able to transmit the thrill of things at thirty than at twenty—just as an actress is much more capable of transmitting the thrill of Juliet at forty than at fourteen. So, let the young teacher not despair if he cannot quite 'get it across' at first. Let him consider, for his encouragement, the life of Siddons.

I have spoken once or twice of 'difficulties' in literature. I do not mean to suggest that there are no difficulties or that there is no need for explanations All I want to convey is that the explanation of difficulties is the housemaid's work of literature, and not the artist's work. Poetry, if we like to think so, is a form of music. Such lines as these:

> O world! O life! O time!
> On whose last steps I climb,
> Trembling at that where I had stood before;
> When will return the glory of your prime?
> No more—O, never more!

—such lines are a kind of music, an appeal to the emotions, with no more 'meaning' or 'moral' than Chopin's Prelude in E minor. But there is this important difference. A sensitive child could be deeply moved by the Prelude without knowing or caring what is meant by Prelude or E or minor. He is not quite so much at ease with a poem In the Prelude the music is made by the combination of elements called notes, which have, singly, no meaning or significance; in poetry the music is made by the combination of certain elements called words, which have singly certain meanings or associations. Sometimes the words get so battered that they spoil the music, like a broken note in the piano. Sometimes the words acquire associations that force themselves incongruously on the hearer and spoil the music, like incongruous noises from without spoiling a beautiful performance. Thus, to take a simple illustration, the once poetic word 'balmy' has now become impossible in a serious piece. 'Priceless' is almost equally impossible, and it is very difficult to make boys take seriously such a phrase as 'with eyes like carbuncles.' On the other hand some beautiful lines may miss their emotional effect because certain of the word elements have no significance for the hearer. A child can get the thrill of beauty from lines like these:

> The moving Moon went up the sky,
> And nowhere did abide:
> Softly she was going up,
> And a star or two beside;

but he will miss something from lines like these:

> Her beams bemocked the sultry main,
> Like April hoar-frost spread;

or like these:

> Thridding the sombre boskage of the wood
> Toward the morning star;

the emotional effect will be partly lost because some of the elements have no significance for the young hearers. They are blank notes in his mental instrument. To deal with these 'meanings' is one of the teacher's difficulties. Where, as is often the case, the difficulty is merely one of vocabulary, simple equivalents of the unusual words can be written up beforehand and treated as a vocabulary exercise without any reference to the poem. If a boy knows beforehand, or at the moment, the appropriate meaning of the unusual words in such lines as these:

> Make me a willow cabin at your gate,
> And call upon my soul within the house.
> Write loyal cantons of contemned love,
> And sing them loud even in the dead of night;
> Holloa your name to the reverberate hills,
> And make the babbling gossips of the air
> Cry out, "Olivia!"

or these:

> No, this my hand will rather
> The multitudinous seas incarnadine—

the music of some wonderful passages will not be lost to him. What is wanted for immediate use is the bare equivalent meaning. Any elaborate discussion of unusual words must be reserved for another occasion.

A detailed discussion of the language and style of a great writer can be a very delightful and valuable lesson for the older boys. It is the kind of lesson that they get far too rarely, and, as applied to fine prose, the kind of lesson they get almost never. Such lessons are naturally more fruitful when the pupils are older than those we are here considering, but they can and should be given (in due proportion and with judicious care) to any class that is old enough to think about what it reads. Part of the initiation into literature should be a glimpse of the glory of language, a fostered sense and feeling that language is a form of beauty with its roots in the imagination.

βῆ δ' ἀκέων παρα θῖνα πολυφλοίςβοιο θαλάσσης.

At the very beginning of his Homer the young reader is taught to consider the effect of πολυφλοίςβοιο; is he equally taught to consider the effect of 'multitudinous'? Surely we take the beauties of our wonderful language too much for granted!

Through the tender mercy of our God, whereby the day-spring from on high hath visited us,
To give light to them that sit in darkness and in the shadow of death, and to guide our feet into the way of peace.

Do we ever stay to dwell on the loveliness of a passage like that, in which almost every word is a poem, or do we not just go passively through it as the hustled tourist goes through a foreign picture-gallery, encompassed with beauty and seeing almost nothing?

Now since these dead bones have already outlasted the living ones of Methuselah, and in a yard underground and thin walls of clay, outworn all the strong and spacious buildings above it, and quietly rested under the drums and tramplings of three conquests....

Do we stop to ask ourselves why 'drums and tramplings' and not 'strife and tumult,' or 'struggles and vicissitudes,' or some other pair? If we are to teach literature we must be able both to feel these touches and to communicate them!

But what we must not do, and what I fear we often do, is to mix two different kinds of lesson. We must not turn the reading of a play or a poem into a study of form or language, or we shall be trying to live on two planes at once The explicatory lesson is one thing, and the presentation of a poem something quite different. What pleasure should we get from a performance of the C minor Symphony if the conductor stopped the orchestra at every occurrence of the main theme to expatiate upon the wonderful significance with which Beethoven can invest a simple rhythmic phrase, or from a performance of the B minor Mass if the choir were silenced while someone explained the harmonic effects that make the hushed close of the *Crucifixus* such a wonderful moment? It is delightful to have these beauties of musical language pointed out to us; but not while we are on the emotional plane of a performance. If explanation and performance are put together we get good from neither; but isn't our usual method of dealing with school literature rather like a blend of explanation and performance? It needs must be that explanations come: but woe unto those that foist them into the very utterance of poetry!

And there are some difficulties that we should not even try to explain. Shakespeare abounds in them. When we sometimes slightingly contrast English indifference to Shakespeare with German enthusiasm for him, we forget that German-Shakespeare is written in a language that every German understands, and that English Shakespeare is written in a language that every Englishman does not understand. It is hardly too much to say that to English people without a literary education Shakespeare is written in a foreign language. Some poets (e.g. Wordsworth) are verbally easy, and some poets (e.g. Francis Thompson) are verbally difficult. Shakespeare is not only difficult but archaic as well; and thus he seems doubly unsuitable for young readers. Fortunately, he is saved for the schools by his wonderful power of re-telling a story in dramatic form, and his equally wonderful power of characterisation, and we may add, his incomparable mastery of word-music. Indeed, it is Shakespeare the musician as much as Shakespeare the dramatist to whom we must introduce our pupils. The teacher's business is to give Shakespeare's scenes and characters the best chance of impressing themselves on a class; and his task, therefore, is to remove the impediments. Now extensive explanation will not only not remove the impediments, it will actually add more. How can such a passage as this be 'explained' to boys and girls of fourteen, who, nevertheless, will respond to the marvellous dramatic appeal of *Macbeth?*—

> ......if the assassination
> Could trammel up the consequence, and catch
> With his surcease, success; that but this blow
> Might be the be-all and the end-all; here,
> But here, upon this bank and shoal of time,
> We'ld jump the life to come.

In vocabulary this passage is not specially difficult, but in significance it is almost inexplicable Indeed, as a matter of word-by-word meaning it is utterly inexplicable; the meaning, as in a passage of music, comes, not from the separate sounds, but from the consort of sounds. Shakespeare, of all our poets, is the least capable of a word-by-word explanation. It is not paradox, but simple fact to say that children understand Shakespeare quite well as long as he is not explained to them. But the teacher must understand more than the children; and until he can follow and feel the great and subtle sweeps of phrase with the occasional poise upon some magically-used word, he had better not attempt to read Shakespeare to a class.

What, then, is the unhappy teacher to do? Well, he must do nothing at all with Shakespeare until he is moderately sure that for himself the impediments have all been removed. Shakespeare is not a musician who can be read at sight. For this reason it seems inadvisable that the first reading of a play should be undertaken by young pupils themselves. If Shakespeare were easy, there would be no better way of class reading than an immediate plunge into part-by-part delivery, as he is difficult it is better that the class should get their first impressions from a skilled and understanding reader, as the teacher must be assumed to be. Certainly no pause should ever be made for explanations. Indeed, it is important that children should get their first impressions, not merely of plays, but of all great literature from a good reader. A poem is, in a sense, a musical score, full of difficulties—some scores being much more difficult than others. A child's ability to receive a poem must not (as in the eighties) be measured by his ability to deliver it. We should no more expect an average boy of thirteen to read *Lycidas* at sight than we should expect the average young pianist to play the F minor Ballade at sight. What impresses each immediately is not the beauty, but the difficulty Let the teacher first read the piece to the class, but read it so that the members of the class will want to read it too. A child should as naturally desire to read or recite a piece of word music as to sing a piece of note music.

I hope I may, without offence, beg teachers to use their best discretion in choosing Shakespeare's plays for school use. They must not allow their enthusiasm to reach the height of belief that there is something sacrosanct called Shakespeare, all as indubitably inspired as our grandparents used to think the Bible, with not a word or a sentence questionable. There are, for children, dull passages that are better omitted or hurried through—that tediously protracted dialogue, for instance, between Malcolm and Macduff in Act iv, Sc. iv of *Macbeth*. Between Act ii, Sc. ii of *Hamlet* and Act ii, Sc. ii of *The Merchant of Venice* there is a wide world of difference, and we must not pretend to children that they are equally splendid. And there are plays that are better left untouched. So few can be read in the elementary school course that it seems to me a pity to lavish time upon *The Comedy of Errors* and *The Taming of the Shrew*, to mention two I have known taken, when there is so much that is finer. Enthusiasm for Shakespeare is delightful in school, but I think we may doubt whether a teacher in whose opinion *Twelfth Night* and *The Taming of the Shrew* are both equally Shakespeare and suitable for children is a person who can safely be trusted to deal with Shakespeare at all.

As soon as possible let scenes be acted, and as soon as possible let the whole play be performed. If it is considered necessary that I should offer here a defence of dramatic performances as a part of education, I will say that the drama is an ancient and honoured form of literature that has enlisted the powers of the greatest poets and afforded rational delight to a hundred generations of civilised beings. The sooner a child becomes familiar with the best forms of theatrical amusement the less likely is he to be permanently attracted by the worst. As I have already pointed out, it is an important, though often forgotten, function of education to teach young men and women the rational use of leisure and the best possibilities of decent amusement. Any school activity that contributes to the amenities of existence and intercourse is a necessary and laudable part of the educational system.

The pupils who take part in performances of plays must learn to speak well and move well, to appreciate character and to express emotion becomingly, to be expansive yet restrained, to subordinate the individual to the whole and to play the game, to be resourceful and self-possessed and to overcome or mitigate personal disabilities. It will hardly be suggested that these are negligible accomplishments. Incidentally it has been found that boys or girls usually regarded as stupid, and incapable of learning, have exhibited unsuspected ability in acting and have gained a new interest in themselves and their possibilities. Ability to do something is the first ingredient in self-respect.

The pupils who only look on miss something of all this, but they get a useful sense of participation in a school activity, and they get, too, something that the drama can specially give—the immediate sense of a completed thing, of an artistic whole, with beginning, middle and end. It is unnecessary to dwell upon the educative value of a spectacle that shows in a spirit of poetry and magnanimity, character in action, developing to greatness or lapsing to disaster, triumphing in apparent failure, or failing in apparent success.

Class performances are joyous and instructive adventures. They may range from happy improvisations to a formal show on a special occasion. In their Elizabethan inadequacy of equipment they make an excellent introduction to the conditions of Shakespearean drama. A school performance even with very limited resources can be delightful and profitable to everybody. I saw a very remarkable and admirable performance of *Richard II* given by the boys of a London elementary school—the youngest actor aged ten and the oldest fourteen—the whole preparation of which, from first reading to first performance was accomplished in twelve weeks, without dislocation of the regular school course The dresses were prepared by the parents and teachers in cooperation, and the function had thus a social and friendly side of great importance to the school. Two of the most delightful performances of *As You Like it* and *A Midsummer Night's Dream* I have ever seen (and I have seen Ada Rehan in both) were given by girls and women of sixteen and upwards in attendance at a London Evening Institute. Of course, performances such as these cannot be compared with regular stage performances for technical efficiency, but the spirit of the young amateur is usually finer.

Visits to public performances of plays studied in class are an officially recognised form of educational activity[1]. This is a great privilege, in which remote districts are naturally unable to share,

---

[1] "Were," unfortunately. The position at the moment is that the Law (as represented by the Board of Education) will not allow pupils to pay for themselves, and the Law (as represented by the Ministry of Health) will not allow the Local Authority to pay for them Well, we always knew what the Law was.

but for town schools it is a privilege that has its dangers as well
as its delights. If I could be sure that pupils would see performances like the Hamlet of Forbes-Robertson, or the Portia
of Ellen Terry—if I could merely be sure that they would see
nothing that dishonoured the spirit of Shakespeare, I should urge
upon teachers the fullest employment of their liberty: but we
have to recognise frankly that professional performances may
sometimes be precisely what children ought not to see. Boys and
girls should never be allowed to see the wood-magic of *A Midsummer Night's Dream* destroyed by the protracted clowning of
Bottom, or to find the flower-sweet loveliness of *Twelfth Night*
sullied by extravagant orgies of supposed comic drunkenness.
Better, far, the feebleness and inadequacy of a school performance
than this kind of efficiency. It would be an outrage, if, in the
exercise of a precious liberty, teachers allowed their pupils to get
their first acquaintance with Shakespeare on the stage from performances in which the sweetness of the music is soured, in
which "time is broke and no proportion kept." The power of
surrender to first impressions is one of the gifts of youth; but it
is a gift with its dangers, and teachers must, therefore, recognise
the imperative duty of ensuring that a child's first impressions of
Shakespeare are neither foul nor misshapen. That does not mean
that we must approach Shakespeare in an attitude of artificial
solemnity. Shakespeare must not be made either unnaturally dull
or unnaturally grotesque. He wrote his plays to give immediate
pleasure to a miscellaneous audience, and he resented liberties
with his text. Anything in our treatment that makes Shakespeare dull or distorted is a crime against his spirit—it is "from
the purpose of playing."

It was in no inglorious time of our history that Englishmen
delighted altogether in dance and song and drama, nor were these
pleasures the privilege of a few or a class. It is a legitimate hope
that a rational use of the drama in schools may bring back to
England an unshamed joy in pleasures of the imagination and in
the purposed expression of wholesome and natural feeling.

We must not overlook prose in our school literature course.
The ear for prose is a rarer thing than the ear for poetry. It is
easier to catch the beauty of Shelley than the beauty of Swift.
Our treatment of prose as pure literature need not be different

from our treatment of poetry Poetry is word music with regular rhythm, prose is word music with irregular rhythm. We all agree that passages of verse should be memorised. How many of us ever set passages of prose?

I plead for a greater use of the Bible in our school reading. To borrow a phrase that readers of Oliver Wendell Holmes will recognise, the Bible needs to be 'depolarised.' We must not adopt a special attitude towards the Bible; we must simply consider it, for present purposes, as a great book of English prose published in 1611, just as Shakespeare should mean a great book of English poetry published in 1623. The Bible will be brought into better perspective if it is drawn upon for selections, like any other seventeenth century literature; and I hope editors and publishers will take care that the chosen portions are presented as rationally as ordinary reading matter—for instance, in paragraphs where the matter is in prose, and in stanzas or strophes where the matter is poetry—of course without 'verse' numbers, and with such simple historical comment or setting as will enable the reader to place the characters or events or circumstances in proper relation to time and place. People too often read the Bible as if its persons and events belonged to another world. The men of the middle ages were much nearer the artistic truth of Biblical narrative when they saw in their pictures apostles and patriarchs and evangelists habited like themselves and walking the streets of familiar cities. The Bible offers us almost every quality of fine prose, and I am sure that the once common practice of making children commit passages to memory had a chastening effect on the general ear and literary conscience. We may all be the better for not reading the Bible fanatically and superstitiously; but we are much the worse for not reading the Bible at all. I think it would be hard to treat the usual school or home Bible as a volume of reading-matter—the associations of the black cover or the gilt edges and the double columns will be too strong. Whenever the repulsive-looking school book called 'The Bible' is used, the atmosphere will be tinged with theological controversy and the inspiration of literature will be enfeebled. Present well chosen extracts from the Bible in association with other literature, and the theological associations will not trouble us. With boys of the elementary school age we must not be too austere—we do not expect them to read all Bacon or all Hooker, and we should not expect them to read all St Paul or all Ezekiel. What we want chiefly to do is to break down the tradition that still keeps our wonderful Bible as a book apart to be read only in some unnatural way at certain moments of a seventh of the week.

Teachers will have their own views of how to deal with long prose works, a novel by Dickens for example Plainly, neither teacher nor class can read the whole of *David Copperfield* or *Pickwick* aloud in a single term. It is unfair to protract the reading of any work. The class will do much by silent reading, but occasionally the teacher will read scenes or passages as a treat—if his reading is not a treat he ought not to be a teacher—and occasionally members of the class will be expected to read to the others. Any book that a class finds 'dry' should not be pursued to the bitter end, however sweet the teacher may think it. It is a mistake to force obnoxious books or opinions upon anybody. In fact, the whole idea of compulsion is alien to the world of art. The quality of nurture is not strained. We have to proceed in education, as evolution itself proceeds, upon the principle of conservative innovation—the principle of attempting no more of the unusual than the system can healthily absorb. There is scriptural authority for not casting Jane Austen before those who like—let us say—any female novelist the reader happens to detest. This is certain, that if you make boys read *The Fair Maid of Perth* when they would rather be reading *Ivanhoe* you will make them dislike Scott altogether. To persist with an unpopular work merely because it has been begun is to make a discipline of what should be a delight, and to disallow a rational exercise of the taste we are trying to cultivate. We must be ready to try any adventurous experiment in education; we must be just as ready to scrap our failures. A short work might well be read right through even if it is not specially liked. Thus, to return to poetry, *Coriolanus* is less liked than *Julius Caesar*, but that is no reason why it should not be read as well. I am, personally, against abbreviated editions If boys can read any of *Pickwick* they can read all *Pickwick*; but I should certainly introduce them to the gentle art of skipping, and advise them not to read *The Convict's Return* or *A Madman's Manuscript* or the story about the Queer Client.

Teachers must not let their passion for synchronism drive them to decreeing that all boys must be reading the same page of the same book at the same time. If children are reading naturally, they will be reading not only different pages but different books. There is a time and a place for synchronous reading; but it must not usurp all times and places. Teachers, will, of course, endeavour to advise and direct the boys in their private reading, and make them familiar with the names of books worth looking at when they are older. But let us all beware of too much zeal, and especially of the false austerity whose truer name is humbug. There is a great deal of nonsense talked about the reading of boys —their *Gems* and their *Magnets*, their wild and woolly cowboys and their super-Sherlockian sleuth-hounds, their dashing school-boy heroes and their utterly villainous cads. In the main these slandered 'dreadfuls' are entirely wholesome and almost fiercely moral. The fact that magistrates periodically denounce them from the bench need not surprise us, for the utterances of magistrates upon all matters connected with education are usually beneath contempt. A year's reading of the boy is probably much healthier than a year's reading of the magistrate. The effect of the weekly 'dreadful' on the mind of the boy is much less harmful than the effect of the daily newspaper on the mind of the magistrate. Boys could not find, if they sought for it, literature so deliberately pernicious as the matter in most of our newspapers—matter deliberately pernicious in suggestion and in suppression. If any reader thinks I am exaggerating let him buy a day's newspapers in London and, having considered them, ask himself if one of their main purposes is not to perpetuate animosity, produce mis-understanding, alienate sympathy and create the atmosphere in which disputes can never be adjusted, troubles avoided or wrongs righted. Nothing that the boy reads does this daily evil. The stories produced for boys are vastly better than most of the novels produced for adults. The teacher who has just indignantly con-fiscated a surreptitiously-read 'Nelson Lee,' has probably sent in to Mudie's his weekly requisition for a volume of sentimental lies, romantic sophistications or furtive sexualism—all of which he takes for truth about life. The boys who devour the weekly 'bloods' are the hopeful boys; the hopeless ones are those who are not interested even in Sexton Blake. What we have to do is to take the boy's passion for reading, and enlarge its range. If he is avid of heroes, let us introduce him to D'Artagnan, Athos, Porthos and Aramis; if he is fascinated by detectives, let us introduce him to the Abbé Faria, who, in those memorable conversations with Edmond Dantès proves himself the best, though not quite the first of all 'detectives' in literature. On the other hand, if we really think that we can so elevate the mind of the elementary schoolboy of thirteen that he will make a sacrifice of his weekly periodicals and ask spontaneously for Southey's *History of Brazil* and Buckley's translation of Aeschylus, we had better abandon the attempt to teach literature at all, and take to Economics or Commerce or Engineering or some other faculty that gets down to facts.

And now having defended the penny 'dreadful,' I will go to the other extreme and propose that the boy of fourteen should be introduced to a kind of reading that finds no place in the usual scheme. I want him to make a first acquaintance with philosophy; and, to put the matter in a simple and concrete way I will suggest that he should begin to read Plato. The reader will no doubt consider the proposal absurd, but, instead of arguing with him, I will beg him to glance through the *Crito* and ask himself, first, what there is in it that a Seventh Standard or Fourth Form boy cannot follow; next, whether anything in our present curriculum provides that kind of reading; and next whether that process of steady mental interrogation, that gradual reduction of abstractions, either to thin air or to something actual, is not precisely the kind of education our pupils need and ought to get. From this point of view it is unimportant whether the teacher is by nature a Platonist, or whether (with Samuel Butler in his fine extension of chronology) he puts Plato with Marcus Aurelius among 'the seven humbugs of Christendom.' The student must first find Plato in order to find out Plato; he must examine humbug to make sure that it really is humbug. The process of that examination is the kind of discipline to which the average English mind is never subjected The *Crito* does not exhaust Plato, and Plato does not exhaust the possibilities of philosophy for young readers.

Generally, I may add, for school purposes, English literature means anything good to read extant in English. Whether it was originally written in Chinese, Hebrew, Greek, Latin, Italian, Spanish, French, German, Irish or Welsh, in the fifth century before Christ or the twentieth century after, makes no difference at all. In the world of education there should be no battle of the ancient and modern books, but one great concord of Humanism.

There are those who hesitate to admit literature to the school curriculum at all, because they dread the effect of the heavy-handed, conscientious teacher upon matter so precious. Is not the bloom of some exquisite pieces eternally destroyed for us by the rough usage they had at school? Had we not better, as one pessimist proposes (knowing that literature needs must come), deliberately choose third or fourth rate stuff for teaching purposes, so that the best may still remain unsullied by association with the routine of school and the preparation for examinations? The danger is very real, but I think we must face it, and not run away from it. The plain fact is that every teacher cannot deal with literature and we must not expect every teacher to deal with it. Most teachers, with proper preparation, could give very effective lessons in arithmetic and 'information' subjects; but no quantity of preparation will enable those teachers to make the literature lessons anything but a dismal failure, if they have not a genuine gift of transmissible appreciation Unfortunately most of our teachers, especially the 'trained certificated teachers' who staff the best elementary schools, are so extraordinarily efficient that they do not know their limitations. They are ready to teach anything and really can teach almost anything, and they would resent a suggestion that they should leave literature alone in schools. But a simple test should convince them. Do they themselves habitually turn to the world's best literature for their own daily refreshment, pleasure or consolation? Do they find in the great poetry and prose of the past and the adventurous poetry and prose of the present the real challenging call of soul to soul? If so, they will probably be very safe persons to pass on to their pupils the torch still burning. But if for them literature is a 'subject,' like geography or history, something that has to be studied or 'got up' from a text book, something to be acquired by means of copious annotation and analysis, and laid aside when examinations are over, then I say such teachers have not the ear for literature and must no more attempt to teach it in school than the tone-deaf teacher must attempt to teach music. There is no crime in not having the ear for literature or music; the crime is lacking the ear and persisting in spoiling the song. Some years ago I was engaged in preparing students for the Teachers' Certificate examination. The classes were new, one examination was immediately pending, and the few students who were taking this withdrew from lessons for their last revisions. *A Midsummer Night's Dream* was the prescribed play that year, and most of them had editions specially laden with the kind of information expected to be demanded in examination questions. At the end of one afternoon I heard a student say to another, "Have you got up the list of merits and defects yet?" "No," replied the other, "I did the merits this afternoon, and I'm going to do the defects to-night." It would plainly be a crime to let teachers to whom that sort of thing meant literature be charged with the task of transmitting literature to children But literature will be safe with the genuine enthusiasts; and as for examinations, the best way of preparing for examinations in literature is not to prepare for them. The defence of the literature teacher is that he has abundantly justified his existence Most people, I imagine, can point to a definite day when the glory of literature was first revealed to them, and often the magician has been a teacher. One day a man read something to you. He didn't tell you anything, or teach you anything, he just read something, and you suddenly found that straight in front of you was a door that led to paradise, and the odd thing was that you had not noticed that door till he showed it to you. That is a fanciful way of putting what generally happens. A few, by natural instinct or happy chance, have found their way alone; but most people, I imagine, have owed their induction to literature directly to some person. The book is before you· but till the mysterious voice cries "Tolle, lege!" you do not read, and the conversion does not happen. Without the teacher most children would never so much as begin the approach to literature

As a last word on this subject let me beg teachers to take a sane view of literature. Let us have no pose or affectation about it. Reading Blake to a class is not going to turn boys into saints. In the other parts of our English course we can be certain of accomplishing something; in literature there is merely a chance that we shall do something for somebody, and in that hope we proceed. The end of great literature is truth, and truth, though sometimes exquisite, is often terrible We do not want a cant of literature in the schools.

So far, I have not isolated 'reading aloud' for separate treatment; but that exercise must not be forgotten. Not very long ago reading in elementary schools was officially fixed to mean nothing but the glib parroting of a paragraph from some wretched primer, three of which formed the total reading of a child for a whole year. How the child pattered his paragraph was all that mattered; what he read did not matter at all. Thus the provision of reading matter was determined entirely by the ability of the children in delivery—i.e. all a boy of thirteen was allowed to read for a whole year, was the kind of stuff one was sure he would be able to patter glibly with fewest chances of breakdown; and the primers were manufactured expressly to fit that requirement. The boy read nothing to himself and nothing was ever read to him. The teacher's sole business was to see that his boys 'passed in reading'; and if he was very thorough he made them read their primers again and again, till many boys (as well as the unhappy teacher) actually knew the wretched stuff by heart. Times have changed for the better now—certainly in opportunity.

Reading, as the technical art of audibly delivering printed matter in an agreeable manner, has to be clearly distinguished from reading as the art of extracting personal delight from *Twelfth Night* or *The Ancient Mariner*, or *David Copperfield*, and it must be worked for as a distinct accomplishment—an accomplishment ·like the music and water-colour painting and dancing of the old genteel establishments for young ladies, but vastly more useful, wholesome and pleasure giving.

This is not a manual of method; but there is one technical point about reading that should be mentioned. The element in reading is the phrase, not the word, just as in music the element is the phrase, not the note. Children are kept at reading words long after they should be reading phrases. The stressing of prepositions, conjunctions and other particles, the relentless rigidity of *tempo* and the usual inflexibility of delivery, all arise from the pupils' habit of reading by words With increasing age and practice the young readers should learn to broaden their phrasing and get the natural *rubato* of civilised speech. The breaking up of verse lines into short ejaculations is specially distressing. What we usually hear from children is this sort of thing:

But where/are the galleons/of Spain,

when we ought to hear one sweeping phrase. Such a line as

The human mortals want their winter here,

is artistically one phrase, not two phrases, or three phrases, or seven words, or ten syllables. It will be phrased, of course, as music is phrased, with clear and unmistakable reference to its rhythmic pattern, but it must be one element, not ten elements. The good teacher of music gets the pupil from notes to phrases as soon as possible; the average teacher of reading keeps his pupil far too long among mere words. The verse reading of children is bad, not because it is sing-song, but because it is unphrased and inflexible sing-song. If poetry is not sing-song it is not song.

IV

Conclusion

Do you remember this passage in *The Pilgrim's Progress*—as the pilgrims passed down that valley?

"Now as they were going along and talking, they espied a Boy feeding his Father's Sheep. The Boy was in very mean Cloaths, but of a very fresh and well-favoured Countenance, and as he sate by himself he Sung. Hark, said Mr Greatheart, to what the Shepherd's Boy saith."

Well, it was a very pretty song, about Contentment.

> He that is down need fear no fall,
> He that is low, no Pride:
> He that is humble ever shall
> Have God to be his Guide.

But I care less for its subject than for the song. Though life condemn him to live it through in the Valley of Humiliation, I want to hear the Shepherd Boy singing

QUILLER-COUCH.

THESE, then, are the six main aspects of English regarded as a school subject, (1) Training in Speech, (2) Training in Talk, (3) Training in Listening, (4) Training in Writing, (5) Training in the Use of Books, (6) The Induction to Literature. And here, perhaps, the reader who has been expecting some wildly subversive programme will ask with impatience what there is new and revolutionary in these proposals "Do we not already," he will exclaim, "teach our children to read and write and speak?"

Well, do we? And can they read and write and speak when they leave us? And do their reading and writing and speech stay by them? It is one of our delusions that we teach these things; and the revolutionary programme amounts precisely to a request that we should abandon the delusion and face the facts, especially the fact that, after fifty years of compulsory education, the population of this country is uneducated. "It is very important," says a Report of the Headmasters' Conference (1916) on the Curriculum of the Preparatory Schools, "it is very important to define precisely the aim and content of the study of English, there will be considerable danger of its being quite ineffective, partly through vagueness, partly through exhausting after a term or two the resources of the subject; we have to remember that a very large number of teachers were brought up on a curriculum which practically ignored the subject altogether." Well, in the preceding pages we have defined "precisely the aim and content of the study of English" as far as it relates to children of the Elementary and Preparatory school age; and does anyone still think that there is any "considerable danger" of the subject's being exhausted "after a term or two"? The real danger is not that teachers will exhaust the subject, but that the subject will exhaust the teachers and consume the whole time-table. But no school work is so necessary and, in the real sense, so practical. If insistence on English causes the postponement or readjustment of mathematics and foreign languages (ancient or modern) in our schools, no harm will be done to mathematics or languages. The sight of little boys and girls of seven or eight working on paper printed sets of sums is, to me, a painful and infuriating spectacle. The sight of little boys of seven or eight laboriously grinding at Latin verbs would also infuriate me if I were in the habit of seeing it. Children of this age ought to be playing at shops and cutting things out of paper instead of filling up paper with meaningless figures and words. The whole of the traditional arithmetic syllabus of elementary schools should be readjusted, and it certainly ought to be purged of its artificial and unpractical difficulties. "Reduce 10587397576 square inches to square miles." There is not the faintest justification for making children work sums of that kind. There is neither educational nor practical value in the exercise. The dull inaccurate child will not be made bright or accurate by any quantity of such sums. All that rigmarole of "thirty-and-a-quarter square yards one square pole, forty square poles one rood, four roods one acre, six-hundred-and-forty acres one square-mile," means nothing, and never could mean any-thing to any child, and to keep him grinding at exercises in it is a crime. Let him know from actual allotments or fields or parks or city squares what poles or acres are, but don't waste precious time and energy on puzzles that should be solved by tables or a calculating machine. People rightly talk of certain poems or plays as 'too hard' for children; but they never seem to think that there are measures and quantities too hard for children. It would be interesting if our experimental psychologists could determine exactly the value to the child mind of such exercises as "Find the H.C.F. of

$$3x^3 - 13x^2 + 23x - 21 \text{ and } 6x^3 + x^2 - 44x + 21."$$

And consider the usual type of 'problem' set to children. Here are one or two, taken straight from a class book for young children:

£36. 15s. was to be equally shared among 42 persons. 6 persons took only the amount of their shares, but 9 others declined their shares, preferring that their amounts should be shared equally among the rest. How much did each of the rest receive?

I have enough money to pay either Brown's wages for 12 days or Smith's wages for 18 days. If I employ both Brown and Smith, for how many days can I pay them with the money?

In 10 days 12 workmen earn £31. 10s. 20 labourers earn the same amount in 9 days. (a) How much more does a workman earn in a day than a labourer? (b) What is the ratio of a workman's wages to a labourer's?

I ask teachers to examine their experience honestly and say whether they have detected the least increase in the general or specific intelligence of any child resulting from a course of such sums? Is it not simply the case that the boy who is 'good' at that kind of work continues good, while the boy who is 'fair' continues fair—that, in fact, there has been, and will be, no accession to the intelligence of either? Elementary school-masters who believe that arithmetic will make children sharp are on just the same plane as public school-masters who believe that Latin will make children write English. Such sums as those quoted above are neither tonic nor nutrient. They are simply tests (and not always good tests) for a rough assessment of mathematical capacity; and there is no good to be got from spending an hour every day on tests. The explanation of the strong arithmetical bias of the elementary schools will be found in the pyramidal requirements of the code given on an earlier page. The whole elementary system revolved round the yearly 'examination day,' when every child in every standard was expected to 'pass' in reading, writing and arithmetic—i.e., in things with immediately assessable results; and so the year was spent in elaborate training for the event. The annual examination day has long been abolished, but still the elaborate training goes on; and as the modern tendency has been to drop simple 'rule' sums in favour of 'problems,' the children are not even so accurate as they used to be. Let us once grasp the indisputable fact that daily doses of arithmetic will no more make children bright than daily doses of Easton syrup will make them strong, and we shall take a saner view of the subject. What children of the elementary school age need to know in mathematics is, first, the arithmetic of daily life—the actually used weights and measures of the household, the values of coins, very simple and practical changes of denomination, and the four manipulations—addition, subtraction, multiplication and division—applied to practical and not fancy quantities. It is citizens', not actuaries' arithmetic they need. They must get to understand fractions—farthings and half-pennies and halves of cakes and quarters of apples are fractions that they already understand, and they can understand decimals as soon as they know the place value of integers. Percentages are only a particular case of fractions They should know how to work easy practical areas and volumes (squares and cubes will introduce them at once to indices) and they should get on to equations and the great practical mystery of logarithms as soon as possible. Wherever possible their arithmetical work should be real—it should mean actual weighing, exchanging, measuring, cutting out and so on. At least half of the time now wasted in the purposeless problems and unreal calculations that constitute elementary school arithmetic could be given to English The whole elementary curriculum needs careful and rigorous reconsideration. A brave President of the Board of Education should appoint an even braver committee to prepare a Report on the needful minima of knowledge—the really necessary things a child should know. To keep boys occupied with problems and interminable reductions is just as valuable as letting the engine of a stationary motor-car run free In my childhood I learned tables containing 'firkins' and 'kilderkins.' To this hour I have not the faintest notion what firkins or kilderkins are, or what they measure. I think it must be wine, for surely nothing but wine deserves to be measured by things with such delightful names

This digression into arithmetic is not irrelevant to our subject.

7—2

for the hearts of most teachers in elementary schools—certainly the hearts of the men teachers—are still set upon arithmetic. Do they not always want to give it the biggest share and choicest place in the time-table? They think it is 'practical.' I want to assure them that most of it is fanciful, unreal and unpractical; and that the real 'practical' subject is the English that they leave to odd corners of the curriculum.

The preparatory schools are as dominated by Latin as the elementary schools are by arithmetic. They, too, must put their house in order. A child should begin his first new language, ancient or modern, at about eleven. Four years will have been gained for his English without loss to his Latin, and he will be a happier child.

What school with its 'show' subjects and 'useful knowledge' often means to our victims let us call on Mr Wells to describe:

I remember seeing a picture of Education in some place—I think it was Education, but quite conceivably it represented the Empire teaching her Sons, and I have a strong impression that it was a wall painting upon some public building in Manchester or Birmingham or Glasgow, but very possibly I am mistaken about that. It represented a glorious woman, with a wise and fearless face, stooping over her children, and pointing them to far horizons. The sky displayed the pearly warmth of a summer dawn, and all the painting was marvellously bright as if with the youth and hope of the delicately beautiful children in the foreground. She was telling them, one felt, of the great prospect of life that opened before them, of the splendours of sea and mountain they might travel and see, the joys of skill they might acquire, of effort and the pride of effort, and the devotions and nobilities it was theirs to achieve. Perhaps even she whispered of the warm triumphant mystery of love that comes at last to those who have patience and unblemished hearts . She was reminding them of their great heritage as English children, rulers of more than one-fifth of mankind, of the obligation to do and be the best that such a pride of empire entails of their essential nobility and knighthood, and of the restraint and charities and disciplined strength that is becoming in knights and rulers ...

The education of Mr Polly did not follow this picture very closely. He went for some time to a National School, which was run on severely economical lines to keep down the rates, by a largely untrained staff; he was set sums to do that he did not understand, and that no one made him understand, he was made to read the Catechism and the Bible with

the utmost industry and an entire disregard of punctuation or significance; caused to imitate writing copies and drawing copies; given object-
lessons on sealing wax and silk-worms and potato bugs and ginger and
iron and suchlike things; taught various other subjects his mind refused
to entertain; and afterwards, when he was about twelve, he was jerked
by his parents to 'finish-off' in a private school of dingy aspect and still
dingier pretensions, where there were no object-lessons, and the studies
of Book-keeping and French were pursued (but never effectually over-
taken) under the guidance of an elderly gentleman, who wore a non-
descript gown and took snuff, wrote copperplate, explained nothing, and
used a cane with remarkable dexterity and gusto.

Mr Polly went into the National School at six, and he left the private
school at fourteen, and by that time his mind was in much the same state
that you would be in, dear reader, if you were operated upon for appen-
dicitis by a well-meaning, boldly enterprising, but rather overworked
and underpaid butcher boy, who was superseded towards the climax of
the operation by a left-handed clerk of high principles but intemperate
habits—that is to say, it was in a thorough mess. The nice little curiosities
and willingness of a child were in a jumbled and thwarted condition,
hacked and cut about—the operators had left, so to speak, all their
sponges and ligatures in the mangled confusion—and Mr Polly had lost
much of his natural confidence, so far as figures and sciences and languages
and the possibilities of learning things were concerned. He thought of
the present world no longer as a wonderland of experiences, but as
geography and history, as the repeating of names that were hard to
pronounce, and lists of products and populations and heights and lengths,
and as lists and dates—oh! and Boredom indescribable. He thought of
religion as the recital of more or less incomprehensible words that were
hard to remember, and of the Divinity as of a limitless Being having the
nature of a school-master and making infinite rules, known and unknown,
rules that were always ruthlessly enforced, and with an infinite capacity
for punishment, and, most horrible of all to think of, limitless powers of
espial. (So to the best of his ability he did not think of that unrelenting
eye.) He was uncertain about the spelling and pronunciation of most of
the words in our beautiful but abundant and perplexing tongue—that
especially was a pity, because words attracted him, and under happier
conditions he might have used them well—he was always doubtful
whether it was eight sevens or nine eights that was sixty-three (he knew
no method for settling the difficulty), and he thought the merit of a
drawing consisted in the care with which it was 'lined in.' 'Lining in'
bored him beyond measure.

The picture is still true.

The most serious objection that can be raised to the proposed scheme of education in English is that we haven't the teachers to work it. Possibly; but then we have never asked for them. As our quoted head-masters say, "We have to remember that a very large number of teachers were brought up on a curriculum which practically ignored the subject altogether." We have never insisted that the chief and crowning qualification of an English teacher is ability to teach English. Students come to the elementary schools from the training colleges and give their practice-lessons in arithmetic and history and geography and nature-study —to say nothing of English itself, and the one feature common to nearly all of them is an inability to speak or read decently. To the student, English is merely one of many competing 'subjects,' and not one that matters much, because it isn't 'about something.' History, geography, mathematics, science—these, to the student, are the real 'subjects,' because they give something to get hold of, something to 'get up'; but English—why, everybody knows English! And so it happens that when, in course of time, these students become qualified teachers and enter the elementary schools, the work that lies nearest to their hands and cries aloud for their finest energies and their toughest patience is precisely the work they are least capable of doing. I suppose there isn't an elementary school teacher in the country who does not feel perfectly capable of teaching all the English needed in elementary schools. I wonder how many there really are? I wonder how many could be safely trusted to undertake even the beginning of our programme, the teaching of standard English speech? Certainly not all; perhaps a long way from all. But that need not always be so. Let the paramount claim of English be admitted and teachers will shape themselves and be shaped for the task of teaching it.

Nothing here said must be taken as doubting or denying a great improvement recently in the teaching of English, or as belittling the wonderful work of many gifted and cultivated teachers. Let us recognise this with all due gratitude and praise; but let us recognise just as honestly that these efforts do not represent the general texture of elementary school work, and that, in particular, some men teachers are still satisfied with a very low standard of accomplishment in English.

Education is one complete process, but it may be viewed in three aspects. There is first the education of the intellect, that opens to us the world of fact and knowledge; there is next the education of the will that guides us in the world of morals; and there is next the education of the emotions, that takes us into the world of art, beauty, feeling, expression, spirit—all the *intangibilia* that make the difference between life and death. In our schools we have done something with the first, a little with the second, but almost nothing with the third. This is true, not merely of the elementary schools, but of nearly all schools and places of instruction. The feelings of the Englishman are uneducated. He is ashamed of emotion and regards any approach to it as improper. He is taught, as Gissing says the lower class women are taught, "few of life's graces and none of its serious concerns," and in the world of high feeling he is an embarrassed boor. Let us call as evidence someone who was not an apostle of the emotions or a prophet of culture or an enthusiast for any form of art, but as complete a specimen of pure mind as one could imagine. Thus writes John Stuart Mill, recording his Continental impressions:

Having so little experience of English life, and the few people I knew being mostly such as had public objects, of a large and personally disinterested kind, at heart, I was ignorant of the low moral tone of what in England is called society; the habit of, not indeed professing, but taking for granted in every mode of implication, that conduct is of course always directed towards low and petty objects; the absence of high feelings which manifests itself by sneering depreciation of all demonstrations of them, and by general abstinence (except among a few of the stricter religionists) from professing any high principle of action at all, except in those preordained cases in which such profession is put on as part of the costume and formalities of the occasion.

I could not then know or estimate the difference between this manner of existence, and that of a people like the French, whose faults, if equally real, are at all events different; among whom sentiments, which by comparison at least may be called elevated, are the current coin of human intercourse, both in books and in private life; and though often evaporating in profession are yet kept alive in the nation at large by constant exercise, and stimulated by sympathy, so as to form a living and active part of the existence of great numbers of persons, and to be recognised and understood by all Neither could I then appreciate the general

culture of the understanding, which results from the habitual exercise
of the feelings, and is thus carried down into the most uneducated classes
of several countries on the Continent, in a degree not equalled in
England, among the so-called educated, except where an unusual ten-
derness of conscience leads to an habitual exercise of the intellect on
questions of right and wrong I did not know the way in which, among
the ordinary English, the absence of interest in things of an unselfish
kind, except occasionally in a special thing here and there, and the habit
of not speaking to others, nor much even to themselves, about the things
in which they do feel interest, causes both their feelings and their intellectual faculties to remain undeveloped, or to develop themselves only
in some single and very limited direction, reducing them, considered as
spiritual beings, to a kind of negative existence. All these things I did
not perceive till long afterwards; but I even then felt, though without
stating it clearly to myself, the contrast between the frank sociality and
amiability of French personal intercourse, and the English mode of
existence in which everybody acts as if everybody else, with few or no
exceptions, was either an enemy or a bore.

Many years have passed since Mill wrote that, and his charge
is still true The English, as a nation, are inarticulate, almost
illiterate, and contemptuously indifferent to art and beauty. If
you think that is too hard a saying, walk the streets of London,
or Manchester, or Birmingham, or any town whatsoever, large
or small, see what you see, and hear what you hear, and ask
yourself if you are being afforded the spectacle of an educated
nation with a sense of beauty We possess the greatest dramatist
of the world and the most contemptible theatre in Europe. We
have an almost unequalled heritage of poetry, and we feel
ashamed to confess any knowledge of it. We have a language of
incomparable beauty and we write and talk illiterate drivel These
are not exaggerations. Go into any club and examine the many
weekly illustrated papers, costing a shilling each, and circulating,
presumably, in classes above the lower—the extravagant British
workman does not, I believe, run to the extravagance of shilling
papers. Here is the beginning of one of the regular articles in
one of the shilling weekly papers for the week in which these
lines are written: "Well, B'lov'dest, I wouldn't care to be in
Pearl White's breeches! Umps. Yes! I s'pose this *does* sound
rather boastful. H'ever, it's like this"—etc., all in the same
manner. Consider the moral, intellectual, and emotional level

of the readers postulated by these papers, and you will scarcely be inclined to urge that they are designed for an educated public. As I have already pointed out, the usual denunciation of the elementary school-boy's penny 'bloods' by magistrates and other licensed humorists is very ridiculous. The delinquent boy's reading will compare very favourably with the periodicals, novels and memoirs consumed by the average adult of the middle or upper classes, and his cheap cinema is quite as healthy as the more expensive theatre. It is not the elementary school product that fills the stalls and boxes for the newest and nudest *revues*. It is not the ex-elementary school-boy who seeks out the horrors of our "Grand Guignol" plays, or who goes one hundred and twenty-seven times to "Frilly Bits" and ends by marrying a chorus-girl and in being divorced by her a year later. We have talked expansively about the classics and the public schools, and we have produced (by the neglect of many) a few scholars of extraordinary and exquisite fineness; but into the education of our great middle and upper classes the humanities have never really entered. No attempt has ever yet been made to give the whole English people a humane, creative education in and through the treasures of their own language and literature. The great educational reform now needed is to begin that universal education. English, in the large sense here used, is the one subject that will cover all three aspects of education—intellectual, moral and emotional—and very specially will it cover all that we at present leave naked and barbarous.

It is a scheme of education that will impede no other form of study, but is the needed preliminary to all forms of study. It is proposed, remember, as a foundation not as a completion. As I have already said, elementary education must be elementary. It must begin something, it must not try to finish everything. Upon the foundation of a sound education in English any future fabric of art, language, science, philosophy, commerce or mechanics can be firmly erected Without that foundation nothing can be firmly erected. Once more I beg the reader not to confuse education with the acquisition of knowledge, of which a man may have much and still be uneducated. A boy goes to school, not to get a final stock of information, but to learn how he may go on learning, and to learn that going on is worth while.

A humane education has no material end in view. It aims at making men, not machines; it aims at giving every human creature the fullest development possible to it. Its cardinal doctrine is "the right of every human soul to enter, unhindered except by the limitation of its own powers and desires, into the full spiritual heritage of the race." It aims at giving "the philosophic temper, the gentle judgment, the interest in knowledge and beauty for their own sake" that mark the harmoniously developed man. Humanism is a matter of life, not of a living. We pretend to believe this, but our practice betrays us; for the latest argument in defence of the 'Greats' man is that certain business people prefer him to any other. Would the value of his education be less if they didn't? A whole book has been produced in America to prove that the classics are a sound business proposition. Well, we haven't got quite to that depth here, yet. Some of us still cling to the old belief that there are things in life immeasurable even in dollars. I have in earlier pages denounced the prevalent and pernicious doctrine that elementary education is the process of fitting children to become factory hands or domestic servants. I want to denounce with equal earnestness the prevalent and pernicious doctrine that education is the process of unfitting children to become factory hands or domestic servants. When teachers urge children to study for the sake of getting good positions, do they not realise how they are falsifying the currency of life? To suggest to boys that a clerk is something better than a carpenter, an insurance-agent better than a bricklayer is entirely wrong. It is not the extension of education to all that is socially dangerous, but the belief that education ought to mean a black-coated calling. Yet no people are more frequently guilty than teachers of suggesting that a boy is 'too good' to go into a workshop and ought to go into an office. The County Councillor who recently urged that as ninety per cent. of the elementary school children would have to go into manual labour they did not need a good education is not more dangerous to society than the teachers who openly or tacitly believe that if elementary school children receive a good education they ought not to go into manual labour. Teachers, especially the teachers in elementary schools, are the last persons on earth who can believe that all men are born equal; they should be the last persons on earth to countenance the belief that a manual labourer who is educated is fitted for something better than manual labour. Surely the experiences of the war should have taught us that it is not what a man has to do that degrades him, but what he is, in habit and association. We must get into our minds the vital truth that education is our contribution to the whole twenty-four hours of man, and not merely to the eight or six or five that he sells to an employer. Vocational or professional training, as we have said, may or may not be education; but into the early foundation stages of education the circumstances of occupation must never be allowed to enter. We want the educated boy to rise; but we want him to rise above himself, not above somebody else. If we teach the village boy to read for himself and think for himself, if we give him, not mere instruction or information, but the ability to take a view of things and share in man's spiritual heritage, it is not because we want him to grow up into the village squire, but because we want him to walk

> in glory and in joy
> Following his plough, along the mountain side.

The beginnings of a humane education here advocated will not involve a domestic revolution, or a rearrangement of the social system, or a new scale of moral values, or a preference of one sort of -ocracy or -ism to any other, or an upheaval of any sort. A humane education is a possession in which rich and poor can be equal without disturbance to their material possessions. In a sense it means the abolition of poverty, for can a man be poor who possesses so much? And we can begin this new world to-morrow if we wish. Let us abandon the view that humanism has to be sought in Rome or in Athens. Here, or nowhere, is our Athens. Greek or Latin may be one means to a humane education, but ninety-nine thousand English people will not be humanely educated because one Englishman learns Greek. Let me quote a passage from Sir Frederic Kenyon's address to the Classical Association in January, 1914:

The cause of the classics is the cause of all imaginative associations, of all intellectual interests.. A man will be a better man of business, a better lawyer, a better merchant, a better stock-broker, a less hide-bound politician, if he keeps alive in his soul the love of literature, the interest in things of the intellect, of which the Greek and Latin classics are the spring and perennial source of refreshment.

Sir Frederic's statement comes to this, that such a being as man in such a world as the present is all the better for a humane education, as distinguished from professional training. Well, that is either true or not true. If it is not true, then the cause of the classics is entirely a lost cause. If it is true, then I say there is not a single word of Sir Frederic's claim that may not be true of the Englishman who knows neither Greek nor Latin, French nor Italian, but who has been shown the glory and the grandeur that are England. If (as we hope) to his English he will add the other great languages, why, so much the wider will be his heritage; but we must not assume that without the classics he is a pauper. The necessity of our plea for English is made clear by the fact that *The Times*, in its leading article on Sir Frederic Kenyon's address, assumed that the only alternative to education by means of Greek and Latin is education by means of 'shilling scientific handbooks.' And at the moment of writing comes another justification, for here is a sentence from *Punch's* review of a book on the classics: "*Punch*, who for four-score years has endeavoured to hold the balance fairly between the claims of science and the classics, welcomes this reasoned defence of ancient letters." You see? It is 'science' or 'ancient letters.' There is no room for the language of Molière or of Dante or of Cervantes or of Goethe; and, of course, none for the mere mother tongue! Mr Kipling also testifies against the language of his Empire, for his story, *Regulus*, adopts with unction the usual cant of the classics

But let me now call another classical witness, a great scholar and a great man, from whom I have already borrowed the quoted sentences on an earlier page. Professor Gilbert Murray in his address to the Classical Association on January 8th, 1918, describes thus the enemy whom it is the special glory of the humanities to oppose·

The enemy has no definite name, though in a certain degree we all know him. He who puts always the body before the spirit, the dead before the living, the ἀναγκαῖον before the καλόν; who makes things only in order to sell them; who has forgotten that there is such a thing as truth, and measures the world by advertisement or by money; who daily defiles the beauty that surrounds him and makes vulgar the tragedy, whose innermost religion is the worship of the Lie in his Soul. The Philistine, the vulgarian, the Great Sophist, the passer of base coin for true, he is all about us and, worse, he has his outposts inside us, persecuting our peace, spoiling our sight, confusing our values, making a man's self seem greater than the race and the present thing more important than the eternal From him and his influence we find our escape by means of the Grammata into that calm world of theirs, where stridency and clamour are forgotten in the ancient stillness, where the strong iron is long since rusted and the rocks of granite broken into dust, but the great things of the human spirit still shine like stars pointing Man's way onward to the great triumph or the great tragedy, and even the little things, the beloved and tender and funny and familiar things, beckon across gulfs of death and change with a magic poignancy, the old things that our dead leaders and forefathers loved, *viva adhuc et desiderio pulcriora.*

That is beautifully, greatly said. We know too well that enemy of beauty and truth! It has no definite name, says Gilbert Murray, but Goethe descried it, and named it, in that tribute to Schiller of which Matthew Arnold reminded us:

> Indessen schritt sein Geist gewaltig fort
> Ins Ewige des Wahren, Guten, Schonen,
> Und hinter ihm in wesenlosem Scheine
> Lag, was uns alle bandigt, das Gemeine.

Well, *das Gemeine* has us in thrall now as never before, but is not the power of the enemy due to the folly that has limited the warriors who might destroy it to the few who could utter the Shibboleth of the classics? How is the enemy's growing tyranny to be most effectively fought to-day—by the weapons of Shakespeare and Milton or of Sophocles and Homer? To escape the foul influence must we always go back to the old things and funny things that are foreign and not to the old things and funny things that are native? Is Prometheus less significant in Shelley than in Aeschylus? It is because I know that the power of the evil is so strong and the power of the good as yet so small that I beg the place of honour in the fight for our own great native force—"the illustrious, cardinal, courtly and curial vernacular" of England.

V

Epilogue

FOR unto us a childe is borne, unto us a Sonne is given, and the government
shalbe upon his shoulder.   ISAIAH.

Do I understand—is it credible—that the rates and taxes are at present being
legally used to teach children to bathe?
                   SIR J D. REES, *in the House of Commons*, 1921.

AND now by way of Epilogue I must add what will seem
an anti-climax. I turn from my aspirations and see an actual
elementary school in an area of mean streets and slums. I see
the squalid, frowsy, raucous women (ex-elementary school-girls),
the dulled, unclean, brutalised men (ex-elementary school-boys),
the ragged, screaming, verminous children (present elementary
school-boys and girls)—all the trimmings of the Christian civilisation that has crowned the evolution of the world Is it not
wildly, grotesquely foolish to talk of a humane education for such as
these?—to cast great language and literature before the unclean?
I have lived too long to be an optimist, but I affirm my faith once
more. To doubt the end or to shirk the struggle is to surrender
all these souls to the enemy. No one can stand before an assembled
school of children, even in the London slums, without feeling
deeply sure that these boys and girls are worth fighting for.
Consider: the children are ours for a few precious moments—
do we make the best of the time? We try to do something for
their minds, something, too, for their bodies; but their starved
and perishing souls, what have we ever done for them? Do we
even try to make school a place of joy for them? Children are
like dogs—it takes so little to make them happy! Follow them
in your mind's eye, as they go out of school at the end of the afternoon. We go to our comfortable homes, and they go, whither?
Many to a course of errand getting; many of the boys to odd jobs

at the local shops; many of the girls to the unending task of 'minding the baby.' They have no room to call their own; no place of escape but the street. Their school is shut against them. They cannot do home lessons, for they have no home in which to work. We do not realise what 'indoors' means to the poor town child. Surely, for them, school should be the daily escape into a world of sweetness and light—of space and breadth and vision! Is it always that, I wonder? Is it always even a place of personal kindness? We are apt to forget sometimes, in what a deep and abiding sense we are the children of love Affection is creative; it fashions and moulds us through all the plastic years of childhood, and, in the end, we are at the best what love has made us. Think of your own childhood and then think how little place there is for love and creative affection in those mean streets of teeming houses! I am purposely thinking of schools in the bad areas, not of those in the best, because I want to think of children the furthest removed from humanism. Let no reader unused to the elementary schools imagine that they are all of one type with identical problems to face! There are schools of street boys and schools of home boys; and in the latter you will find children from happy little lower middle class households where the depth of creative affection could not be exceeded by the most favoured homes of England—and what a pity it is that these little households are so often households of one lonely child. No children could be more sanely and happily brought up than some I have known from homes in which the father's wage was very near the poverty line, and from homes in which there was no father at all, but just a brave mother. It is not of such children that I am thinking—though they need thought—they need to be saved from the material and stuffy ideals of the suburbs; I am thinking of those in whose lives love and affection have no part. It is they who have most need of all the humanising influence we can shed upon them They can be humanised by the work of the school and by the personal kindness of the teachers. They can live, for those few precious hours of school, in an atmosphere of humane thought and feeling. We cannot expect them to be human if we do not humanise them; and our present school system does not humanise them. Slums exist because there are slum-souls, because there are souls that would turn a palace into a rookery of slums.

It is the slum-souls that we must strive to save. The vilest thing in the world is cruelty; the sweetest, human loving-kindness. We shall do little in one generation, little in two; and with one dreadful fraction of poor creatures, stunted in mind and body, nothing at all for ever; but from the rest, slowly, certainly, the enemy with all his frightfulness will be driven back. I know that we can never work miracles of transformation in any number of generations; but I know just as surely that we can do much more than we have done as yet.

And think of those others I have mentioned, all the very many children who are not as foul as the dregs—boys and girls who in mind and body are as rich as could be found in the most favoured schools of the most favoured classes. Surely they are worth fighting for! We have done very little for them so far, except to turn their eyes towards 'a better position in life' and to urge them to learn in order to earn, but we can begin. We can make them decently, humanly articulate, and give them access to the world of ideals, of hope, of beauty. By means of their own language and literature we can begin to educate them in the humane qualities—in breadth of view, in depth of thought, in clearness, accuracy and force of expression, in appreciation of beauty, passion, and nobility, and give them a living sense—

> Of Truth, of Grandeur, Beauty, Love and Hope,
> And melancholy Fear subdued by Faith;
> Of blessed consolations in distress;
> Of moral strength and intellectual Power,
> Of joy in widest commonalty spread.

If I were asked to say in one word what it is that a liberal education gives, I should reply, Vision. "Where there is no vision the people perish."

FINIS

PRINTED IN ENGLAND BY J B. PEACE, M A
AT THE CAMBRIDGE UNIVERSITY PRESS

(

Lightning Source UK Ltd.
Milton Keynes UK
UKHW051831200721
387436UK00009B/429